Practical Military Ordnance Identification

Tom Gersbeck, MFS

Chief Warrant Officer-2 (Ret.)
U.S. Marine Corps, EOD

CRC Press
Taylor & Francis Group
Boca Raton London New York

CRC Press is an imprint of the
Taylor & Francis Group, an **informa** business

Cover: Shows a Russian 122mm, Model 9M22U rocket with a high explosive, fragmentation warhead pierced through two shipping boxes containing four 76mm, L25A4, fixed projectiles. In the background, TM57 & TMN46 anti-tank landmines, and MON100 anti-personnel landmines are staged, prior to disposal.

CRC Press
Taylor & Francis Group
6000 Broken Sound Parkway NW, Suite 300
Boca Raton, FL 33487-2742

© 2014 by Taylor & Francis Group, LLC
CRC Press is an imprint of Taylor & Francis Group, an Informa business

No claim to original U.S. Government works

Printed in Canada on acid-free paper
Version Date: 20150424

International Standard Book Number-13: 978-1-4398-5058-9 (Paperback)

Visit the Taylor & Francis Web site at
http://www.taylorandfrancis.com

and the CRC Press Web site at
http://www.crcpress.com

Practical Military Ordnance Identification

CRC SERIES IN
**PRACTICAL ASPECTS OF CRIMINAL
AND FORENSIC INVESTIGATIONS**

VERNON J. GEBERTH, BBA, MPS, FBINA *Series Editor*

Fourth Edition
Robert R. Hazelwood and Ann Wolbert Burgess

Bloodstain Pattern Analysis: With an Introduction to Crime Scene Reconstruction, Third Edition
Tom Bevel and Ross M. Gardner

Tire Tread and Tire Track Evidence: Recovery and Forensic Examination
William J. Bodziak

Officer-Involved Shootings and Use of Force: Practical Investigative Techniques, Second Edition
David E. Hatch and Randy Dickson

Informants and Undercover Investigations: A Practical Guide to Law, Policy, and Procedure
Dennis G. Fitzgerald

Practical Drug Enforcement, Third Edition
Michael D. Lyman

Cold Case Homicides: Practical Investigative Techniques
Richard H. Walton

Practical Homicide Investigation: Tactics, Procedures, and Forensic Techniques, Fourth Edition
Vernon J. Geberth

Practical Analysis and Reconstruction of Shooting Incidents
Edward E. Hueske

Principles of Bloodstain Pattern Analysis: Theory and Practice
Stuart James, Paul Kish, and T. Paulette Sutton

Global Drug Enforcement: Practical Investigative Techniques
Gregory D. Lee

Practical Investigation of Sex Crimes: A Strategic and Operational Approach
Thomas P. Carney

Principles of Kinesic Interview and Interrogation, Second Edition
Stan Walters

Practical Criminal Investigations in Correctional Facilities
William R. Bell

Practical Aspects of Interview and Interrogation, Second Edition
David E. Zulawski and Douglas E. Wicklander

Forensic Pathology, Second Edition
Dominick J. Di Maio and Vincent J. M. Di Maio

The Practical Methodology of Forensic Photography, Second Edition
David R. Redsicker

Many countries have memorials for their bomb technicians, and in the United States there are two. Located next to the military Explosive Ordnance Disposal (EOD) School on Eglin Air Force Base in Florida, the EOD Memorial carries the names of 300 EOD technicians killed while conducting EOD operations. Outside the Hazardous Device School (HDS) on Redstone Arsenal in Alabama is the memorial for 15 certified public safety bomb (CPSB) technicians killed in the line of duty.

This book is dedicated to all the bomb technicians who have paid the ultimate price.

Contents

List of Illustrations

Series Note

This textbook is part of a series entitled "Practical Aspects of Criminal and Forensic Investigations." This series was created by Vernon J. Geberth, a retired New York City Police Department lieutenant commander, who is an author, educator, and consultant on homicide and forensic investigations.

This series has been designed to provide contemporary, comprehensive, and pragmatic information to the practitioner involved in criminal and forensic investigations by authors who are nationally recognized experts in their respective fields.

Preface

The subject of military ordnance is vast and this book is in no way an attempt at providing an encyclopedic dictionary. What it does provide is a practical deductive process that can be successfully applied to collecting the information required to research relevant publications and to identify military ordnance properly.

In order to provide a quick reference that may assist with following the format of this book, the logic trees in Appendix A can be pulled out. The chapters follow the logic trees and offer a quick reference on ordnance categories, groups, and safety precautions. Appendices B, C, D, and E provide easy access to many of the countless abbreviations and acronyms; as well as relevant definitions and measurement conversions associated with military ordnance.

The motivation for writing this book was driven by the fact that military ordnance is commonly recovered in civilian communities. The problem is compounded when a person attempts to research a recovered munition and is quickly overwhelmed by the amount of information available, much of which is incomplete or inaccurate. Technical Manuals (TMs) on military ordnance are the best references available, but these are updated periodically to ensure accuracy; the version available online or from other nonmilitary sources may not include updates.

The intent of this book is to provide the reader with a foundation on which to build his or her knowledge in this area. It is not a magic bullet, but with a perpetual learning curve associated with this topic, a solid foundation is an absolute requirement. The goal of this book is to offer readers a means of identifying a potential threat and conveying critical information to the authorities who can assist them in resolving a potentially dangerous situation.

Acknowledgments

Professional mentorship lights the path to success in all fields and I have been fortunate to receive guidance from the best. It is impossible to name everyone who has influenced the contents of this book, but special thanks go out to my first EOD team OIC and NCOIC Pete Lawson (Maj. U.S.M.C. Ret.) and Jimmie Alldredge (MGySgt. U.S.M.C. Ret.), who started me off in the right direction, followed by Rick St Amand (MGySgt. U.S.M.C. Ret.), who kept me on track.

After this, my professional development was steered by Kelvin Dudenhoffer (Capt. U.S.M.C. Ret.), Don Smith (Capt. U.S.M.C. Ret.), Russ Hart (Maj. U.S.M.C. Ret.), Steve Negahnquet (Lt.Col. U.S.M.C. Ret.), John Little (Maj. U.S.M.C. Ret.), Doug Finn (Lt.Col. U.S.M.C. Ret.), and Ralph Way, PhD (Lt.Col. U.S.M.C. Ret.), to name a few. Every one of these officers was mentored as an enlisted Marine and warrant officer before moving to the regular officer ranks, when all of them paid-it-forward by mentoring the next generation.

After I retired from the Marine Corps, this mentored guidance continued in the field of forensics. I befriended and then sought and received tutelage from Dr. Neil Haskell, Dr. Patrick Jones, and Roy Crawford, PE.

Dr. Doug Scott introduced me to forensic archeology in a way that shaped the practical process structure outlined in the book.

I am deeply indebted to Dr. Larry Babits, who took the time to provide guidance and insight and mentor me through the process of researching and completing this book. Tom Conte (Senior Chief, Navy Ret.) Robert "Bing" Crosby (Senior Chief, Navy Ret.) and Mark Ladd (BCM Navy Ret.) provided technical insight and photographs, which were extremely helpful, in the underwater ordnance chapter. Bob Hayworth (CWO U.S.M.C. Ret.) provided historical insight as only he could.

John Pope (Capt. Egg Harbor Township, NJ, PD, Ret.), Joe Sweeney (formerly with Pittsburgh PD), Mike Walsh (NYPD Ret.), Tom Lynch (Philadelphia PD Ret.), Dave Gazzara (Atlantic Co NJ Prosecutor's Office), and Greg Everett (LASD) provided tremendous insight on the law enforcement perspective associated with military ordnance encountered in civilian communities.

The photographs provided throughout the book would not have been possible without the support of the EOD community. The access provided to ordnance training aids and libraries and the questions answered by the Marine Corps Base EOD Team in Twenty-Nine Palms, the 2nd EOD Company under 8th Engineer Support Battalion in Camp Lejeune, and the Navy EOD team at NOS Earle allowed this work to come together. I cannot name all of the EOD technicians who provided assistance, but want specifically to thank Maj. Tim Callahan, Capt. William Volz, MGySgt. Dave Alexander, GySgt. Eric Gonzalez, SSgt. Johnny Morris, and SSgt. Evers, who also provided photographs and historical details on a number of munitions.

I want to thank my parents, Edward and Ellen Gersbeck, who raised me with the belief that I could accomplish anything.

To my wife, Laura—without your support this book would have never happened. I love you.

To my kids, Amelia, Kyle, Nicole, and Lauren—seeing you grow, develop, and find your own paths in life makes me very proud of each of you.

The Author

Tom Gersbeck served in the U.S. Marine Corps EOD field while also pursuing an academic education; he retired in 2001 as a chief warrant officer with a masters of forensic sciences degree. Tom has served with the Federal Air Marshal Service (FAMS) as an Explosives Security Specialist and as the Explosives Branch Chief of the National Explosives Detection Canine Teams Program (NEDCTP) before deploying in support of Operation Enduring Freedom and U.S. Department of State programs. Deployments include three to Afghanistan to lead Triage Laboratories in two combined explosives exploitation cell (CEXC) facilities and as the project manager of Task Force Paladin's C-IED mobile training teams operating throughout the country. He was deployed twice in support of the Department of State, once as an operational EOD team leader in Basra, Iraq, and as an advisor to the Tanzanian Peoples Defense Force after the Gongo La Mboto ammunition storage site disaster in February 2011.

Tom ran the International Unexploded Ordnance Training Program at Texas A&M for two years and spent six years with Fairleigh Dickinson University as an adjunct professor. He is an active member of the International Association of Bomb Technicians and Investigators (IABTI), the American Academy of Forensic Sciences (AAFS) and continues to work in his field.

Introduction

> There were 21 U.S. Army personnel killed and 53 injured during Operation Desert Storm as a direct result of handling UXO.
>
> **FM 21-16, UXO Procedures**

War is an enduring component of world history. It is a violent clash of hostile, irreconcilable forces set on dominating or completely destroying the other, and military ordnance are the tools used to accomplish this. Every ordnance item is designed to address a tactical purpose, which is not limited to exploding. Ordnance is extremely dangerous and a worst-case scenario involves deployed munitions that did not "function as designed."

For centuries, many of the brightest engineers, chemists, physicists, mathematicians, and those from other professional fields developed weapons and ordnance for their countries' military. Some believed weapons and ordnance research to be a civic duty, while others argued that a person should not use his or her professional skill to do or enable harm. Nevertheless, countries capable of exploiting the technological development of weapon systems—as well as the ordnance they deliver—had significant tactical and political advantage over those who could not or did not. The history and shape of every nation have been formed by the achievements and failures of these talented people. Yet, the actual hazards associated with military ordnance remain something of a mystery to all but a small group of people who specialize in this area.

The principles of ordnance construction are a scientific and engineering endeavor, where the accurate identification of an unknown munition is a developed skill and in many ways an art. When a munition of unknown origin is inspected, a largely subjective and deductive process based on the elimination of what the item is not and the experience of the examiner is applied. For professionals responsible for protecting the public as well as those in a position that increases the likelihood of their discovering a munition, it is essential that they understand the threats posed by military ordnance.

Ordnance can be deployed in four general ways: thrown, dropped, projected, or placed. Ordnance that was deployed but failed to function as designed is considered Unexploded Ordnance (UXO). Munitions damaged by explosions, fire, or other insults are extremely dangerous and unpredictable;

these should also be treated as UXO due to the unknown internal conditions. A munition or a recognizable component of a munition removed from a current or former war zone area is considered a Remnant of War (RW)—oftentimes referred to as an Explosive Remnant of War (ERW).

Covering every type or design of military ordnance in a single book is impossible and this text is not designed to answer every question or cover all ordnance designs. The focus of this work is on the smaller ordnance items that are more commonly recovered outside military control. The fact that large bombs, torpedoes, truck-mounted missiles, and sea mines exist will certainly be mentioned, but these are not the focus of this book. Additionally, information relevant to ordnance from all countries is presented, not just those manufactured in the United States.

In order to cover a tremendous amount of information and dispel some of the mystery associated with ordnance, this book provides the basic fundamentals of a practical deductive process used to identify unknown military ordnance. Once identified, appropriate safety precautions can be applied to minimize associated hazards. In order to address this topic in a concise manner, the focus will be on identifiable construction features associated with how a munition is designed to function. Though far from absolute, such details offer a measure of constants often found on ordnance. Proper identification and adherence to appropriate safety precautions will allow the application of knowledge-based decisions to protect those directly involved as well as the community at large.

This book is written for public safety bomb technicians, SWAT personnel, Explosives Detection Canine (EDC) handlers, emergency management personnel, beach and park patrol units, forensic laboratory staff, Evidence Response Teams (ERTs), UXO technicians, deminers, Coast Guard personnel, archeologists, all military personnel and other first responders, as well as history enthusiasts, museum employees, and those studying these fields. The easy-to-follow, step-by-step means of applying a practical deductive process to identifying ordnance outlined in this book was written with these professionals in mind.

When military ordnance is recovered in the United States responsibility falls on the military. Whether the munition was manufactured in the United States or another country and found its way into the United States is irrelevant. Circumventing the military's responsibility is not suggested, nor is it an objective of this book. Providing accurate information to those who may encounter them so that they can inform the military EOD experts who can provide assistance is the ultimate goal.

Ordnance is recovered in just about every country. The Los Angeles Sherriff's Department (LASD) offers a good example of the scope of this problem in the United States. The LASD bomb squad responds to an average of 500 explosives-related calls each year, of which approximately 200 (40%) involve military ordnance. In situations such as this, the lack of concise reference materials may result in public servants making decisions without access to insightful information. Conversely, proper application of processes outlined in this text will allow those involved to remain safe, accurately identify threats, and convey reliable information to the nearest military EOD unit.

In order to clarify who is an expert in this area, the two types of bomb technicians recognized in the United States are briefly covered: (1) military Explosive Ordnance Disposal (EOD) technicians, and (2) Certified Public Safety Bomb technicians (CPSB).

Prior to the development of the CPSB program, there were a number of bomb squads operating independently. One of the oldest, continually manned bomb squads in the world is the New York City Police Department bomb squad, which celebrated its 100th year of operation in 2003. In 1941, the first formal US training program was started by the military to train EOD technicians who continue to cover all US military installations and operations worldwide. When requested, the military can provide EOD support to other federal agencies as well as jurisdictions without CPSB assets. They are required to support all ordnance-related recoveries, as the ordnance is military property from "cradle to grave." The CPSB technician's area of responsibility throughout all 50 states and some US territories is controlled by the Department of Justice (DOJ), but it can also be influenced by state and local authorities.

When discussing ordnance, the difference in training and information access warrants mentioning. By definition, both types of bomb technicians are experts in the field of items containing explosives as they have more training and knowledge on this topic than the average person. However, military EOD is a "joint" program, meaning that all four service branches attend the same initial training, share ordnance technologies, employ the same ordnance-related tools, and refer to the same classified publications that are not released to DOJ or CPSB technicians. All EOD personnel from the Army, Navy, Air Force, and Marine Corps attend the basic Naval School Explosive Ordnance Disposal (NAVSCHEOD) course, which is 8 months in length. Navy personnel are also required to attend dive school, as well as two additional months of training on underwater ordnance.

CPSB technicians are also part of a joint program as all CPSB technicians attend the 6-week Hazardous Devices School (HDS) run by the Federal Bureau of Investigation (FBI). The basic bomb squad tools are also standardized by the FBI, but the majority of state and local agencies' equipment depends largely on what their agency can purchase or what they can obtain through grants. Most importantly, when discussing military ordnance, these state and local squads do not have access to the classified military EOD publication system.

Common sources of military munitions recovered or displayed in public areas are:

- Veterans returning with war souvenirs including RWs, ERWs, and UXOs, as many do not realize the hazards associated with these "souvenirs."
- Military and civilian research personnel take ordnance home. This does not happen often, but when it does, positive identification may be impossible if the munition was never fielded.
- Ordnance was buried for disposal, which was a common practice until the 1960s.
- Those developing lands formerly used as military training facilities often locate unknown burial sites.
- Ordnance has been deployed over water for training or combat and munitions have been disposed of at sea, which is still practiced. Such ordnance is commonly pulled up by those in the fisheries industry or washed ashore during storms.
- Ordnance accidently discarded or lost by the military.
- Ordnance recovered from shipwrecks.
- Munitions illegally purchased and transported into the United States for criminal purposes.
- Functional munitions stolen from the military.
- Illegal recovery or theft from military installations. For example, there is a cottage industry, referred to as "scrapping," where scrappers sneak onto impact areas to recover aluminum, copper, and other materials to sell. Many UXO items are also taken accidently or purposefully to sell or use in conjunction with criminal activities.
- Civil War era ordnance that was lost, discarded, or deployed on numerous US battlefields.
- Archeological or museum display: Information on legal use and the display of military ordnance can be found in DA PAM 385-64, Chapter 13-6. When displayed ordnance is inspected by experts, it is often found to contain hazardous components or be fully functional.
- Other: An example of the catch-all "other" is practice hand grenade bodies commonly sold as cigarette lighters and other novelty devices. Minor modifications can make these grenades functional and represent significant hazards. However, it is important to note that once a munition has been modified, information concerning how it was designed to function is no longer valid and it is classified as an improvised explosive device (IED).

What Is Not Covered

This book will not address related topics such as weapon systems, tactical countermeasures, and protective measures, as well as other pyrotechnic and explosive hazards that are not classified as ordnance, such as the explosive components associated with ejecting ordnance from an aircraft. Other issues not covered include the Department of Defense (DoD) Hazard Classification System, transportation considerations, the Posse Comitatus Act, penal codes or other legal consequences, as well as ordnance data sheets or any classified information associated with ordnance. Information addressing many of these issues can be found in the references.

The focus of this book is the application of a practical deductive process to identify unknown ordnance items that are commonly recovered outside military control.

Overview of Energetics Associated with Ordnance

1

Unless you try to do something beyond what you have already mastered, you will never grow.

Ralph Waldo Emerson

Introduction

Understanding how ordnance operates and functions requires a basic understanding of energetic materials, including explosives, propellants, pyrotechnics, pyrophorics, and other reactive substances. These materials are used extensively in military ordnance to provide a number of different effects, including the arming and firing of fuzing systems, deployment of payloads, energizing power sources for fuzing, steerable fins, and guidance systems. These materials also propel ordnance, illuminate the night, allow munitions to be tracked, and stabilize or alter trajectories during flight, as well as serving other functional purposes.

This chapter is broken into six sections that cover relevant characteristics and effects produced by energetics used in military ordnance. The six sections are:

1.1 Terms and Definitions
1.2 High and Low Explosives
1.3 High-Explosive Groups
1.4 Explosive Effects
1.5 Propellants
1.6 Pyrotechnics, Incendiaries, Pyrophorics, and Other Smoke Compounds

Section 1.1: Terms and Definitions

For clarity, a few terms and definitions that apply to more than one section in this chapter and an overview on black powder are offered. Additional definitions and information on energetics is provided in Appendices B, C, and D as well as the Bibliography.

1

Function as designed (FAD): This term is used to define when or how a munition or a component of a munition functions correctly, thus producing the designed effect.

Hygroscopicity: This is the ability or inability of a material to absorb moisture through submersion or absorption. The introduction of moisture into a reactive material may reduce its sensitivity or affect its stability. For example, most pyrotechnic compounds are hygroscopic. When a deployed pyrotechnic candle fails to ignite and impacts the ground, the result is numerous internal cracks where moisture can collect. If this happens and the pyrotechnic material is ignited, the introduction of moisture can result in a mechanical explosion dispersing burning candle fragments.

Reactivity (in relation to compatibility): Energetics that react with other materials, such as various metals, are not desirable for use in military ordnance; however, they have been used in the past.

Sensitivity: How an energetic material reacts to heat, shock, or friction.

Stability: How an energetic material degrades over time. Most military-grade energetics decompose slowly, even under extreme conditions. Decomposition may cause an energetic material to become more or less sensitive, depending on the material and the cause of decomposition.

Toxicity: Most energetics are toxic to some degree and present inhalation, absorption, and ingestion hazards. Always check the Material Safety Data Sheets (MSDS) for all energetic materials encountered.

Black powder: Discovered in the 1100s or even earlier, black powder is the longest continuously used military explosive. Conventional black powder mixtures contain 75% potassium nitrate, 15% charcoal, and 10% sulfur, or a sodium nitrate, coal, and sulfur mixture in slightly different proportions. The "form" or shape of black powder can range from a fine powder to grains over half an inch in diameter. In older munitions, black powder was used as the propellant, fuze firing mixture, booster, and main charge. Though not used as comprehensively today, black powder is still used extensively for pyrotechnic delay fuzes, self-destruct delay elements, expelling charges, bursting charges, and other applications.

Black powder is extremely sensitive to sparks and static electricity; it is also very hygroscopic. With black powder, the burn rate can exceed 1,200 feet per second (fps), but this speed is subject to many variables. When stored in a sealed container, the powder will remain stable for extremely long periods of time. For example, in 2008 and again in 2009, men attempting to inert ordnance from the American Civil War (1861–1865) were killed when drilling into artillery projectiles recovered from battlefields.

Section 1.2: High and Low Explosives

An explosion is a sudden release of energy. There are three types of explosions associated with military ordnance: chemical, mechanical, and nuclear. Only two (chemical and mechanical) will be discussed. A chemical explosion is a reaction resulting in a solid or liquid rapidly changing into gases having a much greater volume than the original substance. A mechanical explosion involves a buildup of pressure within a closed container resulting in an energetic breach of the container.

High Explosive (HE): An explosive material capable of detonating when properly initiated, it will do this whether confined or unconfined. A detonation is the propagation of a self-sustaining shockwave and reaction zone that moves through an explosive material at a velocity greater than the speed of sound in that material, which is the explosive's velocity of detonation. For example, TNT is a commonly used baseline for gauging the performance of other explosives. With a velocity of detonation reaching 22,100 fps, TNT is classified as an HE.

Low Explosive (LE): This is an explosive material that does not detonate as the speed of the reaction through the explosive material is slower than the speed of sound and a self-sustaining shockwave does not form. Low explosives may burn, deflagrate, or explode when unconfined, but are more likely to explode when confined. Deflagration is a fuel-rich explosion that produces intense heat in the form of a fast moving fireball with explosive force. The most common application of low explosives in ordnance are propellants in rocket and missile motors, as ejection charges, or as fuzing delays. To avoid confusion, unless specifically stated, LE will be associated with propellants throughout this text. Additional information on propellants is offered in Section 5 of this chapter. An example of a low explosive is black powder, which has an explosive velocity reaching 1600 fps, thus classifying it as an LE.

High-Explosive Performance Characteristics

Many of the performance characteristics of an explosive material are determined by the speed at which they react. A self-sustaining shockwave moving through an explosive material faster than the speed of sound in the reacting material is a detonation, and this reaction has a measurable velocity. The initial speed at which the shockwave moves is referred to as the Velocity of Detonation (VD). A second important variable is the VD that can be maintained without diminishing in strength, which is known as the Stable Velocity of Detonation (SVD). Both VD and SVD are measured in feet per second or meters per second (mps). Each high-explosive material has a VD and SVD range that is affected by explosive density, temperature, geometry, and method of initiation. Additionally, explosives with a fast VD and consistent SVD are considered "brisant" due to the strong shattering effect produced by the higher speeds.

The initiation of both high and low explosives results in the sudden release of energy—meaning that they both explode. But only high explosives reach the speed required to generate the self-sustaining shockwave consistent with a detonation. When high-explosive-filled munitions are initiated, it is necessary to be familiar with a few terms:

1. High-order detonation: This term is used to describe high-explosive-filled munitions that function as designed. The munition is said to have "high ordered."
2. Sympathetic detonation: If a second munition containing high explosives is close enough to the initial one, the detonation may transmit enough shock through the munition bodies and air gap to initiate a subsequent munition causing it to high order.
3. Low-order detonation: Not to be confused with low explosives, a low-order detonation is when a high-explosive-filled munition is initiated to function as designed, but the result is a partial initiation. The munition is said to have "low ordered." Common causes of a low-order detonation include an understrength, defective, deteriorated, or damaged initiator, booster, or main charge that hinders an explosive material from reaching its VD or maintaining its SVD. A low-order detonation results in a dangerous situation as partially detonated, burnt, and otherwise damaged pieces of explosive materials and other munition components will litter the area.

Section 1.3: High-Explosive Groups

In order for a fuze and a munition to function as designed, explosive components are aligned from the most to least sensitive and least to most powerful. This alignment or configuration is called an explosive train. Explosive trains are a fundamental aspect associated with the use of commercial and military explosives. In military ordnance, this fundamental is used as a means of building safeties into fuzing designs.

To support each step of an explosive train, high explosives are grouped according to their use and explosive characteristics. There are three high-explosive groups used with military ordnance: primary, secondary, and main charge explosives.

Note: Appendix C lists some explosives commonly used with ordnance as well as a few of the energetic properties and characteristics that make them suitable for this application.

1. **Primary/initiating explosives** are the most sensitive and least powerful of the three groups. They are easily initiated by heat, shock, or friction. Primary explosives are used before boosters, which are the next component in explosive trains, because they provide enough energy to reliably initiate less sensitive secondary explosives contained in them. Primary explosives are used in primers, detonators, initiators, leads, relays, and delays that can be initiated by the **heat** of an electrical source or flame, the **shock** of mechanical impact, or the **friction** of explosive compression.

2. **Secondary explosives** are the most powerful and brisant of the three groups. They are used to amplify the primary source's explosive energy and reliably initiate the less sensitive main charge explosive. Secondary explosives are used in boosters located between the primary and main charge explosives. Depending on the size of the primary and type of main charge, a booster may or may not be necessary. Unconfined secondary explosives usually burn or deflagrate when exposed to the **heat** of a flame. They can withstand much greater **shock** from impact, but will reliably detonate from **friction** produced by a primary explosive's explosive compression.

3. **Bursting/main charge explosives** are mixtures of one or more secondary explosives and other materials formulated to produce specific explosive characteristics and effects. Ultimately, in an explosive train, it is the main charge that accomplishes the intended results. Unconfined main charge explosives usually burn or deflagrate when exposed to the **heat** of a flame, can withstand greater impact shock than secondary explosives, and will reliably detonate from the **friction** of explosive compression produced by a very strong primary or secondary explosive (Figure 1.1).

Section 1.4: Explosives' Effects

There are numerous effects produced by military ordnance. Only seven specific effects associated with explosive-filled ordnance will be discussed throughout this text. It is important to note that ordnance design, configuration, materials used in construction, method of deployment, type of explosive filler, and the target itself all influence the application of these effects. Additionally, it is impossible for a single munition to produce all seven effects:

1. **Fragmentation (frag) effect:** Material projected away from the point of detonation at high velocities. There are three types of fragmentation:
a. **Primary frag:** Is produced by the warhead, body, or outer case of the munition. Additional enhancements, such as internal serrations, specially designed fragmentation liners, or external fragmentation

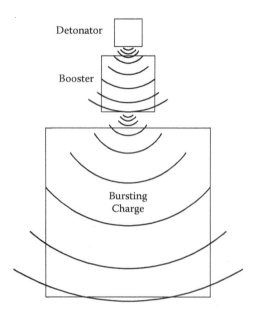

Figure 1.1 Basic HE firing train. Depending on design, the booster in the center may not be required. (From U.S. military technical manual [TM].)

sleeves, greatly increase this effect. A careful balance between the chemistry of the explosive filler and the variables associated with the physical design of a munition produces specific fragmentation sizes and velocities to provide the desired effects on the targets the munition is intended to destroy.

The manipulation of physical design features such as munition shape, case thickness, metal hardness, and additional enhancements influences the potential of a munition. But when filled with an explosive that provides the correct energetic output, the maximum effectiveness of a munition is achieved. Explosives with a high VD tend to produce smaller, jagged fragments moving at high velocities. Explosives with a slower VD usually produce larger, slower moving fragments.

    b. **Shrapnel:** Named after British artillery officer Henry Shrapnel. This term originally defined a specific artillery shell designed in the mid-1780s. There are a few design variations, but examples of common shrapnel projectiles include:
        (1) A cannonball filled with an explosive charge and additional shrapnel (see Figure 4.21 in Chapter 4).
        (2) A projectile with a Powder Train Time Fuze (PTTF) and a flash-tube running down the center to the explosive

Figure 1.2 Shrapnel shell. (Author's photograph.)

charge in the base. The base of the munition is strongly constructed and the body contains preformed individual fragments in front of the explosive charge. This configuration allows the munition to function while in flight, projecting the enhanced frag in a forward direction. This effect is much like a flying shotgun cartridge that explodes, dispersing its contents when it is closer to the target (Figure 1.2). Chapter 4 covers these munitions in greater detail.

c. **Secondary frag:** Is produced by a number of sources and can be as lethal as primary frag. Secondary frag includes nonexplosive components of the munition such as guidance sections and fins, as well as objects from the target or environment such as glass, rock, and wood splinters (Figure 1.3).

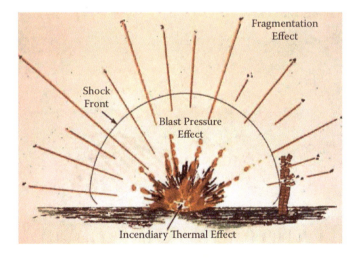

Figure 1.3 Explosive effects. (From U.S. military TM.)

2. **Blast:** A complex interaction of many variables, but primarily involves the **shock front** traveling outward from the point of detonation. The shock front pushes the surrounding environment outward and is followed by **blast pressure** or **overpressure**.

   The effects of a blast in open air, enclosed spaces, underground, and underwater are very different. For example, when the shock front impacts a surface, it reflects causing a **Mach Stem,** which describes the interaction of the shock front, reflected shock front, and converging overpressures (shown in Figure 1.4 as the Mach Stem region). The Mach Stem effect results in almost doubling of the pressures for a short distance off the reflecting surface. As this is a consistent effect, some ordnance items are specifically designed to maximize the extreme overpressures generated by Mach stem effects for specific applications.

3. **Incendiary effect:** Unlike the dramatic fuel-enriched fireballs common in movies, the incendiary effect of a high-explosive detonation is a "quick flash" that can be missed in the blink of an eye. In military ordnance this effect is often enhanced with red or white phosphorus, and metals such as zirconium, aluminum, and magnesium. Section 1.6 of this chapter contains additional information on this topic.

4. **The Munroe effect—shaped charge or hollow charge:** There are few absolutes associated with explosives; however, one effect that is consistent is the "flat-surface concept." In short, when applied against a flat surface, explosive force is consistently focused in the direction of the flat surface. When two flat surfaces are aligned in a converged linear shape, or a continuously converging conical shape (cone), a

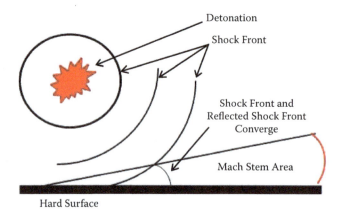

Figure 1.4 Mach stem effect. (Author's graphic.)

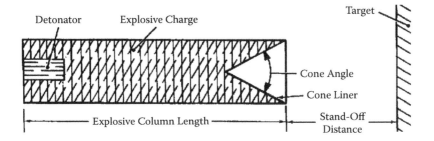

Figure 1.5 Shaped charge configuration. (From U.S. military TM.)

tremendous amount of pressure is focused on a very small area, resulting in a high-performance jet capable of penetrating armor with small amounts of explosive. After jet formation, the remnants of the cone material will form into a slug. After penetrating or missing the target, the slug can travel for many kilometers (Figures 1.5 and 1.6).

This consistently repeatable effect was discovered in 1888 by Charles Munroe, a scientist at the US Naval Academy, and is thus known as the Munroe effect; it is also referred to as a shaped or hollow charge. Capable of penetrating armor, this configuration was quickly applied to ordnance designs. Shaped charges are used extensively in military ordnance; any warhead or munition containing one is "grouped" as a High-Explosive Anti-Tank (HEAT) munition.

A properly constructed conical shaped charge can penetrate armor over six times thicker than the charge diameter. For a shaped charge to produce its intended effect, four factors must be present at the moment of initiation.

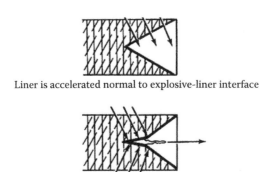

Figure 1.6 Shaped charge jet formation. (From U.S. military TM.)

1. **Explosives:** Brisant explosives are required to generate the greatest results.
2. **Point of initiation:** Conical shaped charges must be initiated at their base on the center axis over the cone to function as designed. If the charge is initiated from any other point, the jet will not be as effective or it will not form at all (Figure 1.5).
3. **Standoff distance:** Upon initiation, a shaped charge must be a specific distance from the target, thus allowing the jet to form for optimal performance. If it is too close or too far from the target, penetration will be degraded or unproductive (Figure 1.5).
4. **Cone specifications: liner material and angle:**
   a. **Orientation:** The open end of the cone must be orientated directly toward the intended target.
   b. **Liner:** The cone liner must be a material that is pliable at high temperatures. While copper is a common liner material, glass and other materials are also effective.
   c. **Angle:** The angle of the cone apex is very important, but may change depending on the intended target and explosive used for the main charge. Shallow angled cones provide less penetration as they move more slowly and use more material from the liner, but the trade-off is that they produce larger entry holes. While steeper angled cones produce smaller entry holes, move faster, and use less material from the liner, they also provide deeper penetration. A 42° angle is considered optimum, providing a speed and liner mass ratio that offers maximum penetration against many target materials (Figure 1.5 and Figure 1.6).

5. **The Miznay–Schardin effect—Explosively Formed Projectile (EFP):** The fundamental flat-surface concept of energy focus also applies to the Miznay–Schardin effect. However, the focus of energetic force associated with a flat surface is further enhanced when this surface is concaved slightly inward as shown in Figure 1.7, and initiated on the center axis of the base over the concaved plate. The result is the formation of an EFP with both the velocity and mass capable of transferring a tremendous amount of energy against a target. Miznay–Schardin effects are greatly enhanced when brisant explosives are used. After penetrating or missing a target an EFP can travel for many kilometers. Other than landmines, EFPs have limited use in military ordnance because the extended standoff requirements limit delivery options. However, this manipulation of energy has many other practical uses, such as the Russian nonelectric blasting cap in Figure 1.8. This initiator

Figure 1.7 Explosively formed projectile (EFP). (From U.S. military TM.)

design is common in Russian grenades and, gram for gram, it is a more effective configuration than the standard American blasting cap with a flat end.

6. **Craters and camouflets:** When an explosion takes place on the surface, a shallow crater is formed. A subsurface explosion capable of breaching the surface will form a "true crater." An understrength subsurface explosion will result in a void or "camouflet" under the surface that is difficult to detect and offers collapse and toxic environment hazards (Figure 1.9). Munitions designed to produce a

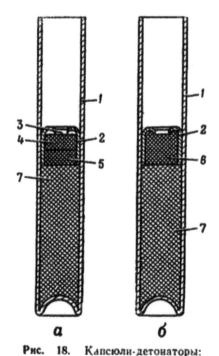

Рис. 18. Капсюли-детонаторы:
а — № 8-А; б — № 8-М; 1 — гильза; 2 —
чашечка; 3 — сетка; 4 — тенерес; 5 —
азид свинца; 6 — гремучая ртуть; 7 —
тетрил (тэн или гексоген)

Figure 1.8 Russian nonelectric blasting cap. (From Russian military TM.)

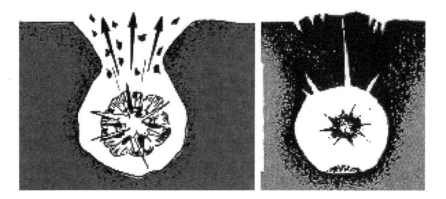

Figure 1.9 True crater (left) and camouflet (right). (From U.S. military TM.)

camouflet are target specific and complex, resulting in additional hazards when they fail to function. An interesting munition designed to produce a camouflet is the French "Durandal."

7. **The Spalling effect** is an effect used to defeat armor without penetrating it. When armor is impacted by a fast moving, dense object or when an explosive is placed in contact with it and detonated, the metal is impacted, compressed, and pushed away from the point of impact or detonation, causing an intense shock wave to move through the armor. If strong enough, when the shock wave abruptly stops on the back side, interior metal flakes off and continues traveling away from the energy source. These metal fragments or flakes are called "spall." This effect can occur on other materials such as concrete or glass. A nonexplosive example of spall is a window impacted by a BB incapable of fully penetrating the glass.

Section 1.5: Propellants

In most cases, propellants react to the same stimuli as HEs, but do so at slower rates more consistent with LEs. Many propellants contain HE with stabilizers and other materials designed to burn at a specific rate, resulting in an energetic burn that produces thrust. There are many materials and chemical mixtures used to propel munitions, dispense ordnance payloads, or move items such as aircraft ejection seats propelled by under-seat rocket motor packs. Propellants can be solids or liquids. Solid propellants, such as smokeless powders, are used to fire or propel projectiles. Liquid propellants are more commonly used in large rocket motors and underwater ordnance.

Many of the performance characteristics of a propellant are determined by the **burn rate** and **ballistic potential,** which are the propellant equivalent of the SVD used to characterize HE material. The burn rate is the speed at which the reaction zone progresses through or consumes a propellant; the ballistic potential is expressed as the total quantity of gas produced when the propellant burns. The performance characteristics, specific formulations, and configuration of a propellant are then used to complete the design factors that maximize munition efficiency.

Most explosives used by the military decompose slowly, even under extreme conditions, but propellants containing nitrocellulose and other volatile materials tend to decompose quickly. If dropped, solid propellants are also prone to damage that can cause malfunctions where liquid propellants are prone to leakage. Any ordnance-related propellant that is damaged or outside its designed configuration is capable of inadvertent initiation.

Low-Explosive Trains

In order for a propellant to function as designed, components are aligned from the most to least sensitive and least to most powerful. In a low-explosive train, this is the means by which a flame is amplified to ignite larger quantities of propellant (Figure 1.10).

a. **Primer or squib** is the smallest, but most sensitive component in the train. Nonelectric primers function upon impact. Electric primers or squibs are initiated by electric current. When a primer or squib functions, a small flame is produced and passed on to the igniter.

b. **Igniter** amplifies the flame from the primer or squib to ignite the main propellant charge. **Note** that ammunition from .22 caliber to 20mm in diameter does not contain an igniter.

c. **Propellant** receives the flame from the igniter or squib and functions as designed.

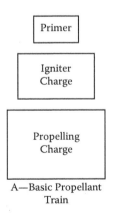

A—Basic Propellant Train

Figure 1.10 Basic propellant firing train. Depending on design, the igniter in the center may not be required. (From U.S. military TM.)

Classes of Propellants

Propellants used in military ordnance are categorized by **class** and **form.** Class refers to the chemical composition of the propellant and form refers to its shape. Due to the complex chemistry and classified nature of many liquid and some solid propellants, this text will only address four general propellant classes: single base, double base, triple base, and composites:

Single-Base Powder (SBP): The primary ingredient in SBP is nitrocellulose, a high explosive with a VD of 24,000 fps. In pure form it is very sensitive to heat, shock, and friction. The addition of other compounds reduces the sensitivity; however, the nitrocellulose content in most SBPs is greater than 85%. Small-diameter SBPs are used in small arms ammunition and have been used in small rocket motors. Some larger "web"-shaped SBPs are used in artillery propellants. Stability for long-term storage of SBP is poor as it is very hygroscopic.

Double-Base Powder (DBP) is composed of nitrocellulose and nitroglycerin. The nitroglycerin is added to gelatinize the nitrocellulose and make it more stable. The mixture increases the energy potential and increases the burning temperature. The increased temperatures wear out larger caliber artillery barrels much faster. For this reason DBP is primarily used in smooth-bore and one-shot weapons, as well as in some rocket motors. Stability for long-term storage of DBP is poor as it is hygroscopic.

Triple-Base Powder (TBP) is a mixture of nitrocellulose, nitroglycerin, and nitroguanidine. Approximately 50% of the mixture is nitroguanidine, which increases the energy potential. With a burning temperature significantly lower than that of nitroglycerin, nitroguanidine-based propellant burns at significantly lower temperatures. The reduction of burning temperatures without sacrificing energy output makes TBP the best propellant for large-caliber artillery. Stability for long-term storage of TBP is poor as it is hygroscopic.

Composite: There are many different combinations that qualify as a composite propellant. Most are composed of a fuel such as aluminum, binders such as synthetic polymers that can also contribute as a fuel, and an oxidizer such as ammonium or potassium perchlorate. Composite propellants are used in rocket, missile, and booster motors. Depending on the materials used, stability for long-term storage of composite propellants varies.

Forms of Propellants

With a known burn rate, the ballistic potential of a propellant can be manipulated by shapes, or "forms," that increase or decrease the surface area as the

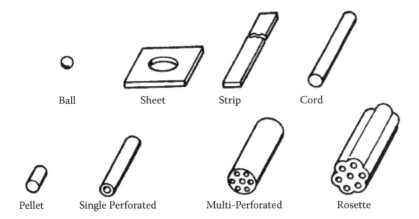

Figure 1.11 Examples of propellant forms. (From U.S. military TM.)

propellant is consumed. The burn rate of any propellant is affected by the heat and pressure generated in the confined space of a motor or chamber. Propellants will burn faster when more surface area is exposed. Propellants come in a variety of forms based on the performance requirements of the munition. Figure 1.11 provides examples of forms, and a brief explanation of how a few different forms work is provided.

Cord form: As these forms burn, the pressure rapidly peaks and then gradually tapers off as the surface area of burning propellant decreases. Classified as a degressive burn, these forms are common in small arms, small rocket motors, and small-caliber artillery.

Single-perforation form: Propellant grains with a single perforation down the center of the cord provide a constant surface area throughout consumption. As the outer surface area decreases, the inner surface area increases, allowing pressure production to remain constant. Classified as a neutral burn, these forms are used in larger rocket motors.

Multiple-perforation and rosette forms: These forms increase the overall surface area as the propellant burns, resulting in initially low pressures that gradually increase as the propellant is consumed. Classified as a progressive burn, these forms are used in large-caliber artillery.

Section 1.6: Pyrotechnics, Incendiaries, Pyrophorics, and Other Smoke Compounds

Each group of military ordnance is designed to address a specific issue or range of tactical problems. Materials that burn are used extensively in ordnance. Uses range from illuminating the night, timing fuze functions, and producing colored smoke to enhancing explosive effects.

Note: In addition to the obvious fire hazards associated with flare compositions, incendiary compounds, and pyrophoric materials, there is also an explosive hazard related to "thermal shock" if water is used to extinguish them. Spraying or dumping water on these materials may result in a mechanical explosion from thermal shock.

Pyrotechnics

There are countless chemical compositions used to provide pyrotechnic effects in ordnance. Common pyrotechnic applications include tracer elements that allow a projectile to be tracked in flight. Flares use pyrotechnic mixtures to produce intense light for illumination or colors for signaling. The color of a burning flare is associated with the type and quantity of metals and intensifiers added to the mixture; the speed at which it burns is determined by the type and quantity of retardants used.

Flare compositions contain fuels, oxidizers, binding agents, retardants, waterproofing materials, and intensifiers for color or smoke production. Common fuels include the powdered aluminum and magnesium with which oxidizers such as sodium nitrate and potassium perchlorate are frequently used. The effectiveness or strength of a composition is determined by its luminous intensity, measured in candlepower.

Incendiary Compounds

These compounds can be employed in liquid, solid, or powder form and are designed to provide a variety of intense incendiary effects. Liquid examples include Napalm, which can contain a hydrocarbon fuel mixed with thickeners, or a mixture of magnesium powder, gasoline, and polyisobutadiene. However, solids are the most common; for example, pellets or liners made of zirconium or magnesium placed near explosives provide an "anti-material" effect when the burning metal is projected from a detonation.

Aluminum powder is commonly blended with high explosives to increase the temperatures generated and oxygen produced during detonation. One of the oldest but lesser known incendiaries is photoflash powder. A mixture of oxidizers and metallic fuels, photoflash powder burns with explosive force when ignited, producing an intense flash of light and high temperatures.

Note: A damaged munition containing photoflash powder in humid or wet environments is extremely dangerous. Photoflash powder reacts with water to produce hydrogen gas and heat, which may initiate the munition.

Pyrophoric Materials

There are a number of applications for pyrophoric materials in ordnance. An example of a multiuse liquid is TriEthylAluminum (TEA), which reacts violently upon contact with water. The energetic diversity of TEA also allows it to be used as a thermobaric explosive as well as a fuel in large rocket motors.

One of the most common pyrophoric materials used in military ordnance is White Phosphorus (WP), which ignites immediately upon contact with oxygen. WP munitions are defined as bursting smokes versus incendiary ordnance because the WP is explosively spread to burn and produce thick white smoke to mark or obscure an area. The smoke is filled with so many particulates that it effectively blocks laser designators used to guide ordnance with pinpoint accuracy. It is important to note that WP will burn until consumed or until the oxygen supply is removed. For example, if burning WP forms a crust and is not exposed to oxygen, it will stop burning. Once this crust is broken, the remaining WP material will immediately reignite when again exposed to oxygen.

Other Smoke Compounds

Colored smoke munitions; defined as burning smokes, pose a fire hazard as they burn to produce smoke. Unlike bursting smokes, the reacting materials remain within the munition as smoke is emitted through vent holes in the body of the munition. Organic dyes or inorganic salts are used to produce differently colored smoke. The most common colors used are red, green, violet, and yellow.

Most white smokes are substantially different from colored smokes in that they use or produce materials that possess severe hazards. Three examples of materials used to produce white smoke are:

1. **Hexachloroethane, aluminum, and zinc oxide:** "HC" is a solid composition that produces gray-white smoke. Upon burning, HC produces zinc chloride and hydrochloric acid that absorb moisture from the air to produce gray-white smoke.
2. **Titanium tetrachloride:** "FM" is a liquid that is mechanically or explosively dispersed, producing dense white smoke upon contact with oxygen. The reaction with oxygen forms hydrochloric acid, which results in the white smoke.
3. **Chlorosulphonic acid:** "FS" is a liquid that is mechanically or explosively dispersed as an aerosol, producing white smoke upon contact with oxygen. The reaction producing the smoke also produces hydrochloric and sulfuric acids.

This inhalation hazard constitutes a "chemical" safety precaution, which will be covered in the next chapter.

The Fundamentals of a Practical Process

2

> Scientific method refers to the body of techniques for investigating phenomena, acquiring new knowledge, or correcting and integrating previous knowledge. It is based on gathering observable, empirical and measureable evidence subject to specific principles of reasoning.
>
> **Sir Isaac Newton, 1687**

Introduction

When attempting to identify unknown ordnance, a scientific method in the form of a practical deductive process is required to interrogate, assess, and identify the item safely. The process outlined in this chapter is the core methodology used by military Explosive Ordnance Disposal (EOD) technicians, but with a more practical science-based application. This proposed process and practical application harmonize well with those applied by archaeologists. An archaeological approach to the recovery of an artifact is a systematic process that starts with an open mind and the development of a general working hypothesis based on what can be observed and all available information. Methodical excavation, careful examination, research of appropriate literature, and a deduction based on the facts and experience of the observer result in constantly evolving, multiple working hypotheses. The goal of this systematic approach is to eliminate as many possibilities as feasible. The remaining choices allow formulation of more specific hypotheses and perhaps an initial conclusion based on the observed characteristics and experience of the evaluator. A successful outcome will result in an accurate identification, professional credibility, additional information, or a better understanding of the past or present. When applied to the identification of military ordnance, the successful outcome of the practical process outlined in this chapter will include everything previously mentioned as well as potentially saving lives.

Throughout this text the dangers and lethality of military ordnance are explained in a manner to preclude readers from minimizing or completely disregarding them. Due to the numerous possible hazards, each munition is considered to contain every possible threat until evidence allows each of them to be eliminated. Unfortunately, the evidence that allows evaluations and decisions to be made is often imprecise because variations in manufacturing techniques and materials used to construct military ordnance ensure that nothing is absolute. The result is a lexicon laced with terms such as "usually," "consistent with," "more often than not," or "as a rule of thumb," followed immediately by "an exception to the rule" and, of course, "an exception to an exception."

Ordnance that has been out of military control may have been fired, damaged, deteriorated, or modified and could possess additional unknown dangers. In an effort to address all the variables, the primary focus of this process is on construction features that help classify the category, group, and type to which the munition belongs.

Category, Group, Type, and Size Definitions

There have been so many different ordnance designs, types, and models made throughout history that no single library in the world is capable of obtaining and storing all the information. In order to classify ordnance, a system of categories, groups, and types was established to provide a basic catalog structure. Accurately cataloging an unidentified ordnance item provides a tremendous amount of insight into the safety precautions (see step 6 later in this chapter) that will apply to the munition. As every munition has a purpose; if that purpose can be ascertained, then most or all of the hazards associated with it can also be determined. The design and construction of a munition are limited by the rules of physics, chemistry, and engineering, as well as a country's manufacturing capabilities. The results are unique characteristics that can assist in identifying the category and group to which a munition belongs.

Each chapter in this book will cover the shapes, designs, construction features, materials, and safety precautions associated with ordnance categories, groups, and types apply the following definitions:

- **Category:** Defines the means of deployment or intended use of the munition. It is the fundamental class an ordnance item falls under. If deployed, "category" answers the question, "How did it get here?"
- **Group:** Defines the effect the munition is designed to produce upon functioning. If deployed, "group" answers the question, "What was it supposed to do when it got here?"
- **Type:** Defines a specific action a munition is designed to perform. Type is primarily limited to fuzes, missiles, rockets, and a small number of other munitions.

To quantify the size of an ordnance item, the following simplistic definitions will be referenced throughout the text:

- **Small** can be concealed in a hand or pocket.
- **Medium** can be carried by a person, but concealment would be difficult.
- **Large** requires two people to lift and a vehicle to transport.
- **Very large** requires a forklift or crane to lift and truck to transport.

Color Codes and Marking Schemes

Most countries paint ordnance to prevent corrosion and offer a means of
identification. Paint colors and schemes include single colors or a combination of background colors, numbers, letters, words, bands, disks, and squares,
or other symbols and patterns to identify the group to which a munition
belongs. In some instances munitions are partially or entirely anodized.

Painted markings can be helpful as a means of identification; however,
ordnance can be repainted, and color schemes can be distorted or completely
removed by impact with a target or exposure to the elements. Additionally,
most countries apply slightly or completely different color codes, which may
render the color scheme as inconclusive proof of a munitions identity. For
example, the United States is currently on its third generation of color schemes
in less than 100 years. Figures 2.1 and 2.2 show current markings and color
codes applying to ordnance manufactured in the United States. Figure 2.3
shows Russian markings; which are also used by many other countries. It is

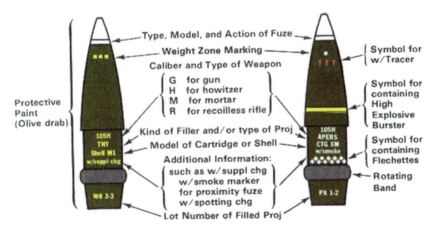

Figure 2.1 Current U.S. projectile marking scheme. (From U.S. military technical manual [TM].)

High explosive		Yellow
Low explosive		Brown
Chemical		Gray
Smoke		Lime-Green
Incendiary		Red
Illumination & Pyrotechnic		White
Armor Defeating		Black
Counter-Measure		Silver
Practice		Blue
Training		Gold

Figure 2.2 U.S. color code chart. (Author's graphic.)

important to note that these color codes do not apply to small arms, blanks, cartridge cases, fuzes, many pyrotechnic devices, and demolition charges.

Stamped Markings

In addition to paint, many ordnance items have markings stamped into the metal as a dependable means of identification. Stamped markings are difficult to cover up or deface and are more apt to survive impact with a target, as well as long-term exposure to the elements (Figure 2.4).

Stamped markings are used by most countries and may include nomenclature, model number, fuze type, serial number, and other pertinent information; however, submunitions, small landmines, and many external and internal components of larger ordnance items may not have markings of any kind. It is also important to note that once outside military control, munitions stamped INERT or EMPTY may have been refilled with hazardous materials.

Seven-Step Practical Process

Ordnance is inherently dangerous; however, adherence to the seven-step practical process outlined next provides a level of safety and greatly increases the probability of an accurate identification. When preparing to approach a munition, always assume that it contains the most hazardous features possible and is in a hazardous condition.

The seven-step practical process for identifying a munition requires an understanding of how ordnance operates and functions, which is discussed throughout the following chapters (Figure 2.5). Depending on the situation, the sequence of these seven steps can change. However, the application of

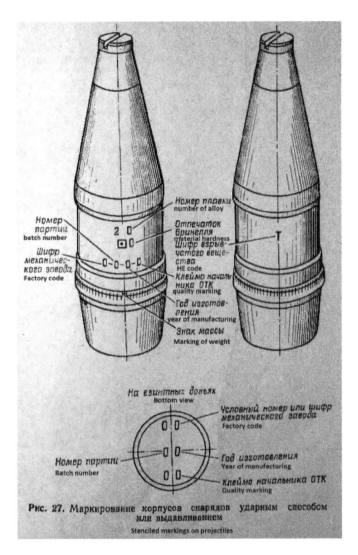

Figure 2.3 Russian projectile marking scheme, which is also applied by many other countries. There is an English translation below each Cyrillic description. (Courtesy of Didzis Jurcins.) Note: The stamped markings on the base of the projectile are not a common practice in the United States and NATO countries.

all seven steps is required to successfully identify and determine the safety requirements associated with a munition. From the moment step 1 begins, steps 2, 3, 4, 5, and 6 are simultaneously addressed. Depending on the munition, some characteristics may be more easily recognized, allowing some steps to be answered quickly. Those remaining unanswered warrant additional consideration during step 7.

Figure 2.4 Russian 100mm high-explosive anti-tank (HEAT) projectile with painted and stamped markings. (Author's photograph.)

Throughout the interrogation process (step 1), do not manually move or touch an ordnance item. Under no circumstances are plungers depressed, vanes rotated, pins removed or replaced, or levers or any other external features moved, as these actions may arm or function a munition.

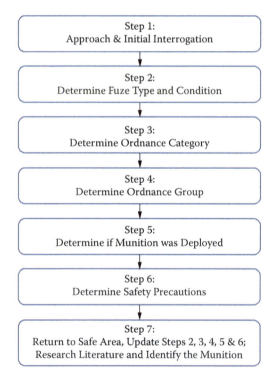

Figure 2.5 The seven-step practical process. (Author's graphic.)

Step 1: Approach and Initial Interrogation

Attempt to identify a munition at a distance with the use of binoculars. If this is impossible and an approach is absolutely required, avoid all venturis and fuze sensing elements. Armed and active or damaged sensing elements may "see" a person approaching, consider him or her a valid target, and function as designed.

Take a pen, paper, calipers (inner and outer), camera (preferably digital), and measuring scale to interrogate the munition and document your findings. Approach the item until it is in view, stop, and begin to determine answers for steps 2–6; continue to address these steps throughout the interrogation process. If the front of the munition can be determined, adjust the route of travel to ensure an initial approach of 45° from the rear of the munition.

After reaching the munition, assess steps 2–5 and adhere to the safety precautions specified in step 6. Begin at one end of the munition and work toward the opposite end, taking note of all identifying construction features. Using the pen, paper, calipers, camera, and measuring scale, document the munition and its location with a rough sketch, measurements, and photography. The single "absolute" associated with ordnance is that every component serves a purpose and provides insight to its identity.

At a minimum, ensure that the length and width of the fuze and munition are documented. Then account for the presence of any identifying features, including fins, rotating bands, venturis, leaking material, color codes, stampings, markings, distinct construction features, and signs of damage, tampering, or modification—as well as damage suggesting that components might be missing. The intent is to document features that will answer steps 2–7. Exit the area via the route taken on approach; return to a safe area.

Note: Proper measurement locations will be addressed later in the text.

Note: Military ordnance, especially practice hand grenade bodies, are often illegally reconfigured to explode and filled with energetic materials. As there is no way of ascertaining the accuracy of unauthorized modifications, it must be assumed that the munition may not function as designed; it is then classified as an improvised explosive device (IED).

Step 2: Determine Fuze Type and Condition (Logic Tree 3, Appendix A)

The fuze constitutes the brains of a munition and there may be two or more present. If maximum consideration is given to ensuring that the fuze does not function, the threat of the munition causing harm is greatly reduced. An external, visible fuze will ease identification; but if the fuze is internal or otherwise covered and cannot be seen, its "type" may still be determined from the category and group (steps 3 and 4) to which the munition belongs. If the munition has been deployed, the fuze is considered to be armed (step 5). If a fuze is damaged or components such as pins or clips have been removed, the fuze is considered armed. If the munition shows signs of alteration or modification, consider the munition armed as the internal configuration is now unknown and may include an alternate fuzing system.

Measurements for the fuze are taken separately from the munition.

Step 3: Determine Ordnance Category (Logic Tree 2, Appendix A)

Category refers to the means of deployment and can be answered by the presence or absence of external features. Ordnance categories and the identifying features that assist in making this determination will be discussed throughout each chapter.

Step 4: Determine Ordnance Group (Logic Tree 2, Appendix A)

Group further defines the munition by identifying its designed effect. Ordnance groups and identifying features that assist in making this determination will be discussed throughout each chapter.

Step 5: Determine if the Munition Was Deployed

If an ordnance item has been thrown, dropped, projected, placed, or otherwise deployed for its intended purpose and failed to function as designed, it is classified as unexploded ordnance (UXO). Another term used to describe UXO is to say it "dud fired" or is a "dud." A munition that was deployed and failed to function is in the most dangerous condition because it should have functioned and the reason it did not is unknown. Additionally, it may have sustained internal damage upon impact with the target and be additionally sensitive.

Indicators to assist in determining if a munition was deployed will be covered for each category.

Step 6: Determine Safety Precautions That Apply to the Munition

There are 16 fundamental safety precautions associated with military ordnance and fuzing systems. There are additional safety precautions related to specific or specialized munitions that are not included in this text. When an unknown piece of ordnance is encountered, all 16 safety precautions are initially adhered to. Throughout the interrogation process, safety precautions associated with categories and groups that can be ruled out are dropped.

Ultimately, there will be fewer possibilities that best match the munition; these safety protocols are adhered to until the situation or munition is taken over by the proper authorities. This deductive process is made easier if the munition's category and group can be ascertained, as many safety precautions are specifically associated with certain categories and groups.

Note: For the submunition, landmine, and underwater ordnance categories, some of the fuzing options and safety precautions will be covered with each group, as the fuze is oftentimes specific to the munition. For all other categories, only the ordnance safety precautions associated with the group will be provided.

Safety precautions are not provided in a specific order because they can vary with different munitions. Consider making a word or phrase to assist in memorizing the safety precautions so that they can be recalled more easily. They are High Explosive (HE), fragmentation (frag), electromagnetic radiation (EMR), static, movement, jet, ejection, chemical, fire, White Phosphorus (WP), Cocked Striker (C/S), Wait Time (W/T), proximity or Variable Time (VT), piezoelectric (PE), Boobytrap (B/T), and influence.

1. High Explosive (HE):
   - Hazard: Explosive blast and overpressure.
   - Safety precautions: Do not expose a munition to heat, shock, or friction. Establish an initial 360° by 300 m exclusion area around the munition. Personnel operating within the blast radius should seek adequate frontal protection. When the amount of explosives is confirmed, increase the exclusion area if necessary.

2. Fragmentation:
   - Hazard: Primary and secondary fragmentation.
   - Safety precautions: Establish an initial 360° by 300 m exclusion area around the munition. Personnel operating within the fragmentation radius should seek adequate frontal and overhead protection. When the amount of explosives is confirmed, increase the exclusion area if necessary.

3. Electromagnetic radiation (EMR):
   - Hazard: The unintentional initiation of an ordnance item or component of a munition; EMR is electrical energy produced by radios, radar, cell phones and other electronic devices. It can affect fuzing and other electronic components, especially if the munition is damaged.
   - Safety precautions: Never use a radio, cell phone, or other electronic device near an unknown ordnance item.

4. Static:
- Hazard: The unintentional initiation of an ordnance item or a component of a munition; static can affect fuzing, other electronic and explosive components, especially if the munition is damaged.
- Safety precautions: When working with ordnance do not wear wool or nylon clothing. Discharge static by placing the back of the hand on dirt or by touching a grounded item.

5. Movement:
- Hazard: The unintentional initiation of an ordnance item or a component of a munition; many fuzes contain free-floating impact or inertia weights, cocked strikers, and other hazards that are extremely sensitive to movement. Additionally, many fuzing systems are completely internal to a munition and cannot be seen. Movement of an armed or damaged munition may cause it to function as designed.
- Safety precautions: Positive identification and condition determination must be made prior to moving any munition.

6. Jet:
- Hazard: Alludes to the energy focus from a munition containing a conical shaped charge or explosively formed projectile (EFP); a shaped charge jet penetrates armor for short distances, but the remaining material from the cone is deformed into a teardrop-shaped "slug" that constitutes a downrange hazard, as some can travel for miles. An EFP also forms a slug that constitutes a substantial downrange hazard. This safety precaution applies to all high-explosive anti-tank (HEAT) munitions and other ordnance containing shaped charges and EFPs.
- Safety precautions: Do not orient a munition toward populated areas. Due to the frequent use of piezoelectric fuzing in HEAT munitions, adhere to PE, EMR, and static safety precautions until the munition is positively identified.

7. Ejection:
- Hazard: Components that are forcibly ejected or locked in place during deployment; examples include the explosive or spring-loaded ejection of submunitions, pyrotechnic candles, fin assemblies, and probes. For ordnance items with motors such as rockets and missiles, ejection applies to the areas in front of the munition and behind a venturi, which may be on the base or sides of some munitions.
- Safety precautions: Initially approach an item from a 45° angle to its rear. Work outside the probable opening deployment arc of

fins, probes, payloads, and other deployable hazards. Never move past venturis or in front of a munition that contains a motor. Personnel operating outside the exclusion area but in the potential path of a rocket or missile should be moved.

8. Chemical:
   - Hazard: Contamination or inhalation of toxic materials such as chemical weapons, riot-control agents, smoke produced by burning pyrotechnics, heavy metals used in guidance systems, toxic propellant mixtures, some smoke mixtures, and explosive main charges such as the chemicals used in fuel air explosive (FAE) munitions.
   - Safety precautions: Establish an immediate exclusion area of 450 meters and a 2,000 meter downwind hazard area.

9. Fire:
   - Hazard: Intense fire capable of spreading quickly; applies to munitions containing pyrophoric, pyrotechnic, and incendiary components or payloads.
   - Safety precautions: Move away from a burning munition in an upwind direction and establish an exclusion area in accordance with the HE safety precaution. Never approach a smoking munition. If an ordnance item is burning, expect a higher order detonation, do not inhale the smoke, do not look directly at burning pyrotechnics, and do not attempt to extinguish burning explosives, pyrophoric materials, or pyrotechnic mixtures as this could result in thermal shock and a detonation.

10. White Phosphorus (WP):
   - Hazard: Differs from the "fire" precaution and specifically applies to munitions containing white or red phosphorus (WP or RP) as the smoke produced by these materials is highly toxic. When ignited, both WP and RP burn until fully consumed unless the oxygen source is cut off by submersion in water or the formation of a crust over the unconsumed material. When this happens, the remaining material will be protected by the water or crust. When removed from the water or if the crust is broken, WP will immediately reignite, but RP will not. If a damaged WP or RP munition is smoking, always expect a high-order detonation as WP burns at temperatures higher than the detonating temperature of most explosive burster charges.
   - Safety precautions: Immediately move away from a smoking or burning WP or RP munition, do not inhale the smoke, and do not disturb crusted-over WP. Apply the fire safety precaution for combustible materials in the area.

11. Cocked Striker (C/S):
   - Hazard: The unintentional initiation of an ordnance item or component of a munition; a C/S is a firing pin under spring tension, held in place by a positive block within the fuzing mechanism. During fuze arming and functioning, the positive block securing the firing pin should have moved, allowing the striker to move and function the fuze, but this process was somehow disrupted. Movement of any kind can enable the munition to function as designed. In some fuzes, the detonator versus the pin is the moving component and is classified as a "cocked detonator." The C/S safety precaution is applied to these configurations as the same functioning principles and hazards apply.
   - Safety precautions: Due to the extreme sensitivity associated with a C/S, do not move or in any way disturb a deployed munition suspected of containing a fuze with a C/S.

12. Wait Time (W/T):
   - Hazard: The unexpected initiation of an ordnance item or component of a munition. Many fuzes and components of ordnance items contain batteries, capacitors, and pyrotechnic and clockwork mechanisms designed to provide time delays ranging from milliseconds to months before functioning the munition. An appropriate W/T is taken prior to approaching a munition with a pyrotechnic, clockwork (C/W), electronic delay, or Self-Destruct (S/D) feature.
   - Safety precautions: Attempt to identify ordnance at a distance with the use of binoculars. If a delay fuze or component is identified, do not approach the munition. Access to appropriate military EOD publications is required to research a W/T for any munition.

13. Proximity or Variable Time (VT):
   - Hazard: The unintentional initiation of an ordnance item or a component of a munition; VT refers to a specific type of fuze containing an electronic sensing element. The sensing element determines distance (proximity) of the munition to a target. This information is then used to function the munition at the desired distance from the target. When located in the nose or front of a munition, these fuzes usually sustain severe damage upon impact. However, some missiles position VT fuzing elements on their body's sides.
   - Safety precautions: Initially approach an item from a 45° angle from its rear. Never move in front of a VT element as it may "sense" you as the intended target and function as designed.

14. Piezoelectric (PE):
   - Hazard: The unintentional initiation of an ordnance item or component of a munition; PE fuzing systems use a quartz crystal to produce electric current when impacted or otherwise stressed. The current produced by the PE crystal is used to fire an electric detonator in the fuze. PE fuzes can retain their ability to function indefinitely and are considered extremely hazardous.
   - Safety precautions: Do not stress the piezoelectric crystal element of a munition in any way. PE fuzing systems are commonly used with HEAT warheads and this safety precaution is applied until its presence is conclusively ruled out.

15. Boobytrap (B/T):
   - Hazard: The unintentional initiation of an ordnance item or component of a munition; B/Ts can be internal or external to a munition, mechanically or electrically functioned, and are always assumed to be present with landmines.
   - Safety precautions: Due to the extreme danger associated with a B/T, do not move or in any way disturb a deployed munition suspected of containing a boobytrap.

16. Influence:
   The influence safety precaution covers magnetic, acoustic, and seismic fuzing systems. Influence fuzing is used on some surface munitions; however, it is most commonly associated with underwater ordnance.
   a. Magnetic:
   - Hazard: The unintentional initiation of an ordnance item or component of a munition; magnetic fuzing senses ferrous metal and functions when specific thresholds are met.
   - Safety precautions: Attempt to identify ordnance at a distance with the use of binoculars. If a magnetic fuzing system is identified, do not approach the munition.
   b. Acoustic:
   - Hazard: The unintentional initiation of an ordnance item or component of a munition; acoustic fuzing senses noise and functions when specific thresholds are met.
   - Safety precautions: Attempt to identify ordnance at a distance with the use of binoculars. If an acoustic fuzing system is identified, do not approach the munition.
   c. Seismic:
   - Hazard: The unintentional initiation of an ordnance item or component of a munition; seismic fuzing senses vibrations in the ground, air, or water and functions when specific thresholds are met.

- Safety precautions: Attempt to identify ordnance at a distance with the use of binoculars. If a seismic fuzing system is identified, do not approach the munition.

Step 7: Identify the Munition

Travel out on the same route taken when approaching the munition, return to a safe area, and research the verifiable information obtained from the munition. Using military manuals, historical ordnance literature, and consultation with military EOD personnel, validate the findings and attempt to conclusively identify the munition.

As there is no single aspect or feature that is absolute and able to provide conclusive identification, all of the information obtained during steps 1–6 must be considered. As there are too many ordnance-related variables for any process to be infallible, even the most experienced practitioner must be cautious when considering a conclusive identification.

In addition to the construction characteristics and configuration of an unknown munition, other considerations that may affect identification include:

- Damage from high-speed impact, fire, and other sources
- Decay-related damage
- Modifications that may have been made after leaving a controlled environment
- If present, color codes and other painted markings are helpful if accurate.

Warning: Colors and other painted markings may have been applied according to unpublished color schemes or with whatever color paint was available.

Cultural aspects of design are helpful indicators and may include unique shapes, symbols, or components such as wooden handles on a grenade.

Closing

The accurate identification of an unknown munition can be a life or death decision, as it will define the safety precautions taken. Ordnance is inherently dangerous, especially if it has been deployed or is old, deteriorated, damaged, or modified in any way. Until proven otherwise, always consider a deployed munition to be armed and in its most hazardous condition.

Fundamentals of Fuze Functioning

3

For 200 years we have been treading so closely in the footsteps of precedent that the ordinary time-fuze of 1855 scarcely differs in principle of applying a composition to graduate and convey a flame to the charge of a shell from that in vogue at Dale in 1632.

Commander John Dahlgren, United States Navy, 1856

Introduction

The statement made by Commander Dahlgren was accurate, but came at a pivotal time in military ordnance development. Until the mid-1800s advancements in ordnance and fuzing system design were minimal as the weapon systems and tactics they supported went largely unchanged; however, the mid-nineteenth century's industrial revolution resulted in advancing new manufacturing methods that were quickly applied to ordnance development. The only requirement to perfect many new ordnance designs was a large-scale testbed. Throughout the 1840s and 1850s numerous European revolutions, the Mexican–American War, and the Crimean War had little impact as the battlefields were far removed from the scientists and engineers involved in ordnance design, as well as manufacturers. The situation changed with the onset of the American Civil War, when the opportunity to develop and immediately test new ordnance designs on a massive scale was realized as the two largest armies in the world clashed on their own soil. Many technological advancements from this period are still evident in munition designs today: most notably, fuze designs focusing on reliability.

The fuze is the brain of a munition, as once deployed, it arms and determines when and how the munition will function. Prior to deployment, a fuze must be in a safe condition so that personnel can handle, transport, and employ it without prematurely functioning. Whether a fuze was designed in the nineteenth or twenty-first century, its effectiveness is associated with three fundamental principles:

1. Functionality: A fuze must contain all essential components required to initiate the ordnance correctly. When properly deployed, the fuze should arm and function as designed.
2. Precision: The fuze must function the munition at precisely the right time to ensure that the munition's maximum effectiveness is achieved.
3. Dependability: Fuzes must have low failure or "dud" rates.

Today's fuzes are highly engineered and function with unprecedented precision, yet many munitions still fail to function correctly. Common causes of malfunctions include improper predeployment preparation, incorrect deployment, disruption during deployment, deterioration of components, improper interaction or impact with the target, and material defects in fuzing components.

Functioning as Designed

As the fuze is the means of properly functioning a munition, it is important to discuss the four distinct phases or conditions a correctly deployed fuze goes through: safe, committed, arm, and final action or function as designed.

Prior to functioning as designed, a fuze must move through the safe and committed phases to "arm." Fuzing designs apply the actions associated with deployment into the arming process for a fuze. Depending on the delivery system, arming can involve the removal of pins or wires; retardation from a sudden and sustained loss of velocity, deceleration "creep," violent acceleration, or centrifugal force induced by spin; a surge of electrical power; or arming vanes that must rotate as the fuze moves through air or water. Generally, arming is a sequence of actions prior to, during, and sometimes after munition delivery that coincides with the method of delivery, fuzing design, and the intended target.

As a fuze leaves the "safe" condition and goes through the arming process, it reaches a point at which this process cannot be stopped; this point is referred to as "committed." Many fuzes require a combination of three distinct actions to go from safe through committed to become armed. Once armed, a munition should "function as designed" or carry out its "final action," thus functioning the munition.

When ordnance does not function when it should have, it is referred to as a "dud." What phase the fuze was in when the process was interrupted will determine the "condition" of the fuze. Conditions include safe, partially armed, armed, or unknown. As ordnance is inherently dangerous, any munition that has been deployed and failed to function, is damaged, or its condition cannot be determined must be considered **armed** and in an extremely hazardous condition.

Warning: Some fuzes are designed to appear benign but contain anti-disturbance features, self-destruct mechanisms, or other hazards when deployed.

The practice of a country copying the designs of another country also makes accurate fuze identification more complicated. Accurate identification is required before a safe course of action can be considered.

Merging Philosophies

One fundamental issue associated with accurately identifying a fuze arises from merging philosophies, which are supported by varying manufacturing capabilities. Many fuzes made in different countries look exactly alike on the outside, but function differently. As mentioned in Chapter 2, when attempting to identify ordnance, everything is important. "Everything" includes construction features, materials used to manufacture it, stamping, markings, paintings, packaging, and all observable characteristics. Some countries apply a commonsense approach to fuze manufacturing and concentrate on proven designs while some do not. For example, consider this quote from 1968:

> The aircraft gun fuzes appear to have been based on WWII German types of proven reliability. In some instances exact copies of internal elements have been noted. In other cases German mechanical principles have been assimilated with Soviet munition philosophy design practices, resulting in a composite but highly effective design. (Communist Block Projected Munitions, Fuzes in Vietnam, September 23, 1968)

Fuze Locations Defined

As with all technical disciplines, there are lexicon differences between specific fields that can become confusing. When discussing ordnance, specifically the fuzing location, often has different terms applied. For example, an air-delivered bomb defines a fuze in the front as a "nose fuze" and a fuze in the rear as a "tail fuze"; a fuze in the side of a bomb is defined as a "transverse fuze" (Figures 3.1 and 3.2). A projectile fired from a howitzer or gun defines a fuze in the front as a "point fuze" and a fuze in the rear as a "base fuze" (Figure 3.3). Fuzes located inside a munition and not externally observable are referred to as "internal" fuzing (See Logic Tree 3, Appendix A).

These terms, though different, define fuzing groups and types and are important to understand. These terms and additional naming convention nuances will be addressed in this and following chapters.

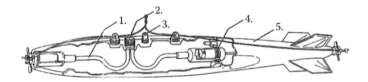

Figure 3.1 (1) Internal electrical plumbing that allows nose and tail fuzing elements to communicate. (2) Charging well for electrical charging. Also connects to both nose and tail fuzing elements. (3) Hoisting lug. (4) Tail fuze. (5) Conical fins. (From U.S. military training manual [TM].)

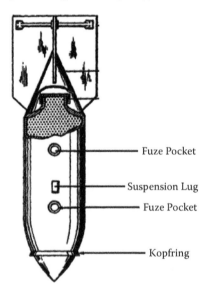

Figure 3.2 Transverse fuzing in a bomb. (From U.S. military TM.)

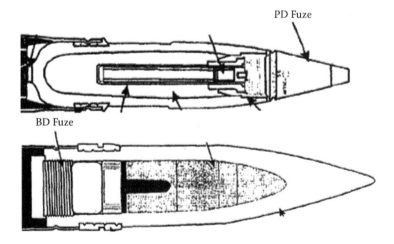

Figure 3.3 Point detonating (PD) and base detonating (BD) fuzing in projectiles. (From U.S. military TM.)

The Seven-Step Practical Process Applied to Fuzes

Whether by itself, externally employed on ordnance where it can be observed, or inside a munition where it cannot be seen, the fuze is always interrogated as a stand-alone threat. By itself, a fuze possesses substantial hazards and many of the physically larger fuzes possess a substantial amount of explosive. If deployed with or inside ordnance, the fuze is capable of functioning the munition and must be accurately identified.

As fuzes can be found separate from or with a munition, they constitute their own category of "fuze." After this, they are broken down into groups and then further defined by "type."

Whether a fuze is interrogated by itself or as part of an assembled munition, it is classified as "fuze." It is then grouped and typed by the manner in which it is designed to function as well as its location in or on the munition. Fuzes are further defined according to their specific functioning design, including whether or not there are multiple ways in which they can function. If the category and group of the ordnance can be determined, many possible fuzing configurations can be eliminated, thus greatly reducing possibilities and increasing the probability of a correct identification.

Warning: If located on or potentially inside a munition, the fuze may be capable of functioning the ordnance during initial approach. Attempt to identify the fuze from a safe distance before approaching any unknown ordnance.

The Seven-Step Practical Process—Fuze

Step 1: Approach and initial interrogation. Attempt to identify the fuze and the munition at a distance. Interrogation: ensure the length and width of the exposed fuze components are documented as well as any identifying features (e.g., material, color, stampings, markings, distinct construction features, and signs of damage, tampering, or modification, as well as damage suggesting components are missing). If observable, information associated with the fuze type, model, settings, or lot number may be stenciled or stamped on the fuze body (Figure 3.4a and 3.4b).

Steps 2, 4, and 5. Determine fuze, group, type, and condition (logic tree 3, Appendix A). A fuze is considered armed and in a hazardous condition if:

1. It is employed on or inside a munition that was deployed (i.e. thrown, dropped, projected, or placed) for its intended purpose.
2. It is missing any arming pins, vanes, clips, etc. that should be present.
3. It has sustained any damage.

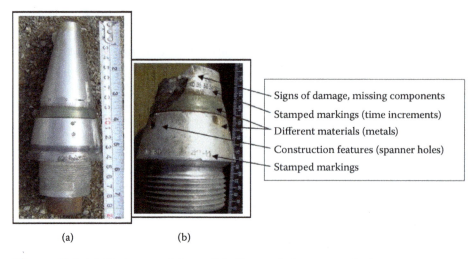

Signs of damage, missing components
Stamped markings (time increments)
Different materials (metals)
Construction features (spanner holes)
Stamped markings

(a)   (b)

Figure 3.4 (a) Undamaged fuze. (b) Signs of damage, missing components; stamped markings (time increments); different materials (metals); construction features (spanner holes); stamped markings. (Author's photograph.)

Step 3: Determine ordnance category (logic tree 3, Appendix A). The category is "fuze."

Step 6: Determine safety precautions that apply to the fuze. The safety precautions that specifically apply to fuzes are high explosive (HE), fragmentation (frag), electromagnetic radiation (EMR), static, movement, ejection, chemical, fire, cocked striker (C/S), wait time (W/T), proximity or variable time (VT), piezoelectric (PE), boobytrap (B/T), and influence.

Note: Do not underestimate the hazards of fuzes due to their relatively small size as they may possess substantial explosive charges.

Step 7: Research the literature and identify the fuze. Due to location, a fuze in the munition's nose or point may sustain substantial damage upon impact, thus making conclusive identification impossible. All internal and unidentifiable fuzes are treated as the most hazardous possibility until proven otherwise.

Fuze Groups and Types

Throughout this chapter, the six basic fuze "groups" and twenty-one "types" that function with distinct differences will be covered. General identification features and safety precautions associated with each group and type will be provided. It is important to note that groups, types, identification features, and safety precautions covered are in no way all encompassing as it is impossible to cover all fuzing nuances. The fuzing groups and types that will be covered are:

1. Impact
  a. Point Detonating (PD)
  b. Base Detonating (BD)
  c. Transverse
  d. Point-Initiating Base-Detonating (PIBD)
    (1) Electrical
    (2) Mechanical
  e. All-way-acting

2. Time
  a. Powder Train Time Fuze (PTTF)
  b. Mechanical Time (MT)
  c. Electronic Time (ET)
  d. Clockwork, short and long delay
  e. Chemical delay

3. Proximity or variable time

4. Pressure
  a. Direct pressure
  b. Pressure/tension release
  c. Hydraulic pressure
  d. Cumulative pressure
  e. Hydrostatic pressure
  f. Contact (underwater)

5. Influence
  a. Magnetic
  b. Acoustic
  c. Seismic

6. Anti-Disturbance (A/D):

Groups

1. Impact fuzes are designed to function upon impact with the target. There are electrical and mechanical versions that can function instantaneously upon impact or after a delay. The delay feature is usually very short but allows a munition to penetrate a target or pass through thick foliage such as treetops before functioning. Depending on design, impact fuzes can be located in the front, rear, side, or inside a munition. There are numerous designs incorporating a variety of mechanical and/or electrical arming actions or sequences. These include pins or wires that must be removed; retardation, setback, or centrifugal force requirements; electrical power; or arming vanes that must be rotated. Hazards associated with these fuzes vary. As with all fuzes, positive identification is required to ensure appropriate safety precautions are adhered to. Five fuze types will be covered under the impact group.

1a. Point detonating, impact, or nose fuzing: Are most commonly used in the nose or front of bombs, projectiles, rifle grenades, rockets, missiles, submunitions, and underwater ordnance. These fuzes are designed to function upon impact with a target. If deployed, positive identification is often difficult due to damage incurred upon impact.

General identification features associated with PD, impact, or nose fuzes include (Figures 3.5 and 3.6):

- **Construction:** Single or multipiece.
- **Materials:** Usually made with steel, aluminum, or Bakelite plastics. Older fuzes and those made for naval ordnance are often made with brass.
- **Markings:** Nomenclature may be stamped or stenciled on observable sections of the fuze.
- **Other:**
  - A common feature is a selector lever resembling a standard screw with "D" and "SQ," for delay and super quick, or other markings relevant to the country of manufacture.

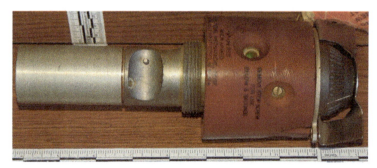

Figure 3.5 M904E4 bomb fuze (nose). (Author's photograph.)

Figure 3.6 Various impact or point detonating (PD) fuzes for projectiles and rockets. (Author's photograph.)

- A firing pin may be enclosed in the forward end under a crimped closure disk.
- The firing pin or striker may protrude from the forward end of the fuze.
- A viewing window displaying a color associated with the condition of the fuze is present on some bomb fuzes.

- **Common employment:** HE and WP munitions.
- **General safety precautions** for PD fuzes include:
  - HE, frag, movement.
  - C/S for some PD fuzes.
  - EMR, static, and W/T for fuzes with electrical components.

Note: A bomb may contain an unseen fuze in the front fuze cavity covered by a penetrator or plug.

1b. Base detonating, and tail fuzing: Are most commonly used in the base or aft-end of bombs, projectiles, rifle grenades, rockets, missiles, sub-munitions, and underwater ordnance, as well as a specific group of hand grenades. These fuzes are designed to function upon impact with a target. Unlike fuzes located in front, fuzes in the rear are protected from direct impact. When deployed, identification is often complicated as many of these fuzes are flush with the munition's base or covered by fins, tracer elements, and other obstructions such as a rocket motor.

Figure 3.7 Three different BD fuzes covered by tracer elements. (Author's photograph.)

General identification features associated with BD, impact, or tail fuzes include (Figure 3.7):

- **Construction:** Consists of a single piece protruding from the munition; however, additional tracer components can make this difficult to determine.
- **Materials:** Usually made with steel, aluminum, or Bakelite plastics. Older fuzes and those made for naval ordnance are often made with brass.
- **Markings:** Nomenclature may be stamped or stenciled on observable sections of the fuze.
- **Other:**
  - Spanner holes in the base of a fuze are flush with the munition body.
  - There are wrench flats on protruding portions of the fuze.
  - A munition has explosive related color codes, but is absent other fuzing.
  - A viewing window displaying a color associated with the condition of the fuze is present on some bomb fuzes. As the colors displayed are specific to each fuze, the appropriate reference will be required to determine the meaning of the color present.

- **Common employment:** Armor-Piercing High Explosive (APHE) and many High-Explosive Anti-Tank (HEAT) munitions.
- **General safety precautions** for BD fuzes include:
  - HE, frag, movement.
  - C/S for some BD fuzes.
  - EMR, static, and W/T for fuzes with electrical components, and PE for those with a piezoelectric crystal.

Note: A bomb may contain an unseen fuze in the base fuze cavity covered by a plug.

1c. Transverse fuzing: Transverse fuzes are located in a munition's side and most commonly employed with bombs, missiles, and underwater ordnance. Unlike nose and tail fuzes, transverse fuzes are usually flush with or countersunk into the side of a munition.

General identification features associated with transverse fuzes include (Figures 3.8 and 3.9):

- **Construction:** Multipiece face with a locking ring.
- **Materials:** Usually made with steel or aluminum. Those made for naval ordnance are often made with brass.

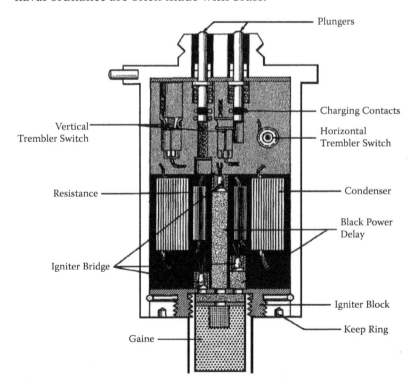

Figure 3.8 German, WWII era impact, electrical, transverse bomb fuze. (From U.S. military TM.)

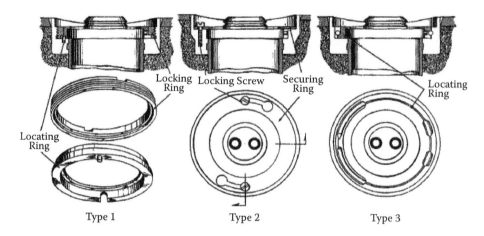

Figure 3.9 Methods of securing transverse bomb fuzing. (From U.S. military TM.)

- **Markings:** Nomenclature may be stamped or stenciled on observable sections of the fuze.
- **Other:**
  - Spanner holes in the visible face of the fuze.
  - Electrical contacts on the visible face of the fuze.

- **Common employment:** Bombs and underwater ordnance.
- **General safety precautions** for transverse fuzes include:
  - HE, frag, movement.
  - B/T or anti-withdrawal feature on some fuzes.
  - EMR, static, and W/T for fuzes with electrical components.

1d. Point-Initiating Base Detonating Fuzing: Consists of two separate fuzing components that must work together to function the munition correctly. A Point Initiating (PI) element in the point or nose of the munition works similarly to a PD fuze, while the BD element in the base or tail of the munition functions in the same manner as a stand-alone BD or tail fuze. PIBD fuzes are designed to function upon impact with a target. There are numerous designs and the PI element may appear to be a PD/nose fuze or be completely covered. When deployed, positive identification of the PI element is often difficult due to damage incurred upon impact. While the BD element in the rear would normally be protected from direct impact, it may be covered by fins, tracer elements, and other obstructions such as a rocket motor, or be completely internal, thus complicating identification. There are two distinctly different PIBD designs that need to be covered separately: (1) electrical PIBD fuzes and (2) mechanical PIBD fuzes.

1. **Electrical PIBD fuzes:** There are two common designs of electrically initiated PIBD fuzes:
   a. A piezoelectric (PE) crystal in the PI element. Upon impact with a target, the PE crystal generates an electrical impulse that travels down a wire or other conductive pathways to an electric detonator in the BD element (Figure 3.10).
   b. A power source connected to two dome-shaped electrical contacts separated by a gap that closes when crushed upon impact with a target; this closes the circuit providing power to an electric detonator in the BD element (Figure 3.11).

Figure 3.10 Cutaway photograph of M409, 152mm, HEAT projectile. (Author's photograph.)

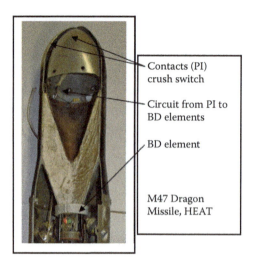

Contacts (PI) crush switch

Circuit from PI to BD elements

BD element

M47 Dragon Missile, HEAT

Figure 3.11 Crush switch. M47 Dragon Missile, HEAT. (Author's photograph.)

2. **Mechanical PIBD fuzes** incorporate a "spitback" action in which the PI element functions upon impact with the target and initiates the BD element by "spitting back" explosive force through a tube or void within the munition. In Figure 3.12, spitback action involves the initiating element (circled in red) shooting a flame or ballistic disk rearward, passing through the center of the warhead, initiating the BD element (circled in blue), and enabling the munition to function as designed. This fuzing design is commonly found in HEAT and older shrapnel munitions.

General identification features associated with PIBD fuzes include:
- **Construction:** A break in the major diameter near the forward end of the projectile body or a standoff spike on the forward end, which is consistent with HEAT warheads.
    - A conical ogive crimped or screwed to the body with no obvious fuze related protrusions.
    - A mechanical PIBD fuze may have a BD element that changes the base of the munition, as shown in Figure 3.13.
- **Materials**: Depending on the munition, can be heavy steel, thin or thick aluminum.
- **Markings:** Nomenclature may be stamped or stenciled on observable sections of the fuze.

- **Common employment:** PIBD fuzes are so common with HEAT warheads from all categories that an electric PIBD fuze must always be suspected on a HEAT warhead until proven otherwise.

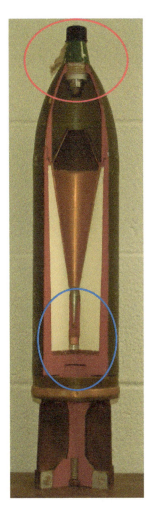

Figure 3.12 The internal configuration of a mechanical PIBD spitback fuze in a BK463UM, 122mm, HEAT projectile. (Author's photograph.)

- **General safety precautions** for *electrical* PIBD fuzes include:
  - HE, frag, movement, PE, EMR, static, and W/T

- **General safety precautions** for *mechanical* PIBD fuzes include:
  - HE, frag, movement

Note: For both electrical and mechanical PIBD fuzes: A common design aspect is a graze-sensitive feature that may include a C/S. A graze-sensitive feature allows a fuze to function upon a glancing impact with the target. Once armed, fuzes containing this characteristic are extremely sensitive.

Figure 3.13 External appearance of a 57mm HE projectile base (left) and the base detonating element of an M307 (right). (Author's photograph.)

1e. All-way-acting fuzing: Is designed to function at all possible angles of impact and used in munitions that are purposefully deployed without stabilization. When a munition is not stabilized in flight, its angle of impact cannot be predetermined, requiring an operationally flexible fuze. When deployed, identification may be difficult, as oftentimes these fuzes are inside the munition. General identification features associated with all-way-acting fuzes include (Figure 3.14):

- **Construction:** Single or multipiece.
- **Materials:** Steel or aluminum.
- **Markings:** Nomenclature may be stamped or stenciled on observable sections of the fuze.
- **Other:**
  - A lack of stabilization devices on a munition such as rotating bands, slanted venturis, fins, ribbons, etc.
  - Mechanical versions use vanes to arm, but vanes may detach during arming process.
  - Electrical versions will have contacts or an electrical connector.
  - Suspect an internal all-way-acting fuze if a munition has explosives-related color codes or markings, but an observable fuze or recognized means of functioning is absent.

- **Common employment:** Submunitions, fire bombs (napalm), and "impact" fuzes for hand grenades.

Figure 3.14 Electrical and mechanical all-way-acting fire bomb fuzes. From left to right: M23A1, WP igniter with M918 *mechanical* all-way-acting fuze; M23A1, WP igniter with FMU-7/B *electrical* all-way-acting fuze; MK273 MOD 0, magnesium Teflon igniter with M918 *mechanical* all-way-acting fuze. (Author's photograph.)

- **General safety precautions** for all-way acting fuzes include:
  - HE, frag, movement.
  - EMR, static, and W/T for fuzes with electrical components.

2. Time fuzes are designed to function a munition during deployment or after reaching its target. There are electrical, mechanical, and pyrotechnic delay designs that can be located in the front, rear, side, or inside the munition. A common application for time fuzes is to function a munition in flight to deploy a payload. There are many different designs incorporating a variety of mechanical and/or electrical arming actions or sequences. These include pins or wires that must be removed; retardation, setback, or centrifugal force requirements; and electrical power or arming vanes that must be rotated. Time delays also serve as self-destruct (S/D) backup features in some fuzes to ensure the munition functions if the primary means fails. For example, antiaircraft artillery (AAA) that is fired and does not hit an aircraft will fall back in close proximity to the forces that fired it and detonate. To ensure this does not happen, an S/D feature is used so that the AAA will detonate at altitude and rain down in small pieces versus an intact munition.

Five fuze types will be covered under the time group.

2a. Powder Train Time Fuzes (PTTFs): Are most commonly found in the nose, but is also located in the base or inside projectiles, rifle grenades, hand grenades, rockets, submunitions, and underwater ordnance. If the fuze can be seen after deployment, positive identification can usually be made as most designs withstand impact well. PTTFs are the oldest type of time fuze

Figure 3.15 Powder train time fuzes (PTTFs). From left: U.S. M84A1; U.S. model 1907; Chinese MS-3A. (Author's photograph.)

containing a black powder train that burns for a predetermined length of time before functioning the next component of the fuze.

General identification features associated with PTTF include (Figures 3.15 and 3.16):

- **Construction:** Multipiece, with at least one brass ring, a time calibration scale, and vent holes or gaps between rings.
- **Materials:** Made entirely of brass or have at least one brass component containing the black powder train.
- **Markings:** Timing increments and nomenclature are usually stamped on observable sections of the fuze.

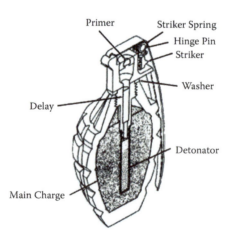

Figure 3.16 U.S. MK-2 fragmentation grenade with a pyrotechnic delay element in the fuze. (From U.S. military TM.)

- **Other:** Vent holes or gaps between rings are necessary to allow smoke and gas to escape as black powder train burns.
- **Common employment:** Dispensers, delay element in hand grenades, self-destruct, or secondary means of functioning a fuze. This is a very common fuze on historical relic ammunition and can still be functional.
- **General safety precautions** for PTTFs include:
- HE, frag, movement, and W/T.

Note: PTTF designs vary; those without an inherent booster or expelling charge will not possess the HE and frag hazards.

Mechanical Time (MT) Fuzes: Are most commonly found in the nose, but are also located in the base or insideprojectiles, rifle grenades, rockets, submunitions, and underwater ordnance. If the fuze can be seen after deployment, positive identification can usually be made as most designs withstand impact well. To increase realiability, many MT fuzes have an impact backup feature displayed as "Super-Quick" or SQ.

General identification features associated with MT fuzes include:

- **Construction:** Multi-piece with a time calibration scale or single piece with a viewing window to see setting (Figures 3.17 and 3.18).
  - Setting lugs or slots to accommodate tools for setting the fuze.
  - Many MT fuzes also incorporate a SQ selector lever.

Figure 3.17 Mechanical time (MT) fuzes for projectiles. From left: U.S. MK 51 MOD 4; Russian VM-30 series; U.S. M577, M564, M772, and M520A1. (Author's photograph.)

Figure 3.18 MK339 MOD1 MT fuze for an air-dropped dispenser. (Author's photograph.)

- **Materials:** Usually made with steel or aluminum.
- **Markings:** Timing increments and nomenclature are usually stamped on observable sections of the fuze. On some fuzes the time calibration scale is in meters allowing a desired distance to be set and the fuze automatically calculates the time.
- **Common Employment:** Dispensers, flechette munitions, Anti-Aircraft Artillery (AAA), self-destruct, or secondary means of initiating a fuze.

- General safety precautions for MT fuzes:
  - HE, Frag, Movement & C/S.
  - All MT fuzes employ a clockwork timing mechanism to release a C/S at the pre-set time.

Note: MT fuze designs vary; those without an inherent booster or expelling charge may not possess the HE and frag hazards.

2c. Electronic time fuzes are most commonly found in the nose of projectiles and rockets, but are also employed with submunitions, rockets, and missiles. Unlike other time fuzes, ET fuzes use electrical power sources to run timing circuits and function an electric detonator. ET fuzes are very versatile; depending on model, time settings can be made manually, programmed with an electronic fuze setter, or be armed remotely. To increase reliability, many ET fuzes have an impact backup feature. After deployment, if the fuze can be seen, positive identification is usually possible due to unique design features. General identification features associated with ET fuzes include (Figure 3.19):

- **Construction:** One or two pieces and may have a time set viewing window on the body.
- **Materials:** Usually an aluminum or composite outer body, and plastic.

Figure 3.19 M767A1 electronic time (ET) fuze with a time calibration window. (Author's photograph.)

- **Markings:** Nomenclature may be stamped on observable sections of the fuze, but a common characteristic of ET fuzes is that they have no external features.
- **Other:** A lack of the time calibration scale, which is common with other time fuzes.
- **Common employment (W/T):** Dispensers.
- **General safety precautions** for ET fuzes include:
- HE, frag, movement, EMR, static, and W/T.

2d. Clockwork (C/W) long-delay fuzes are commonly used in the base or side of bombs and submunitions. Unlike other time fuzes, these fuzes are designed as area denial munitions and commonly employ an anti-withdrawal device such as the one seen in Figure 3.20a. C/W long-delay fuzes use clockwork mechanisms capable of running for hundreds of hours and offer setting options ranging from a few minutes to a few months (Figure 3.20b). To prevent successful explosive ordnance disposal (EOD) countermeasures, most designs include an anti-withdrawal feature that will immediately function the fuze if removal is attempted (Figure 3.20c).

General identification features associated with C/W long-delay fuzes include:

- **Construction:** One or two pieces. Some have a time set viewing window on the body.

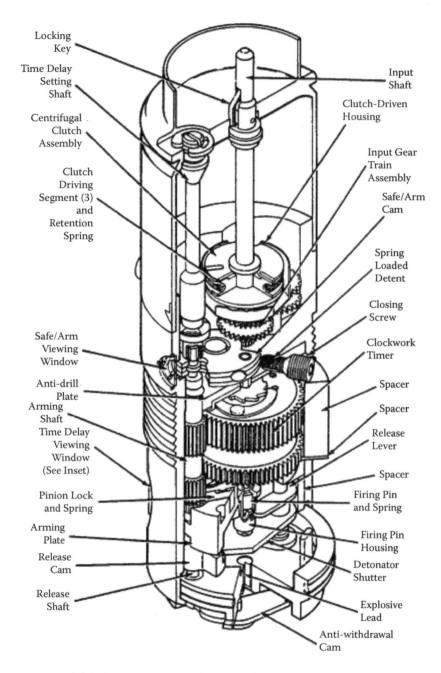

Figure 3.20(a) MK 346 clockwork (C/W) long-delay fuze. Note the "anti-withdrawal cam" near the base. (From U.S. military TM.)

Figure 3.20(b) Cutaway of an MK 346 fuze. Note that the gears and the "window," which displays the time setting, are below the threads, meaning these areas are not accessible when the fuze is installed. (Courtesy of Tom Conte, U.S.N. Ret.)

Figure 3.20(c) The "anti-withdrawal cam" of the MK 346 fuze deployed. The cam deploys into a groove that restricts its removal from the bomb. (Courtesy of Tom Conte, U.S.N. Ret.)

- **Materials:** Usually an aluminum body.
- **Markings:** Nomenclature may be stamped on observable sections of the fuze.
- **Other:** The time delay is set prior to fuze installation and the time set viewing window may not be visible when deployed. These fuzes are often designed to look similar to other fuzes; positive identification is crucial.
- **Common employment:** Bombs and submunitions.
- **General safety precautions** for C/W long-delay fuzes include:
- HE, frag, movement, C/S, and W/T.

2e. Chemical long-delay fuzes: Are commonly used in the base or side of bombs. They employ an acid to degrade the positive block holding a C/S. Different acid ampoules are used to vary fuze functioning times and are quite accurate. The M123 fuze in Figure 3.21 is capable of delays of up to 144 hours. Storage issues and the onset of ET fuzes have rendered these fuzes

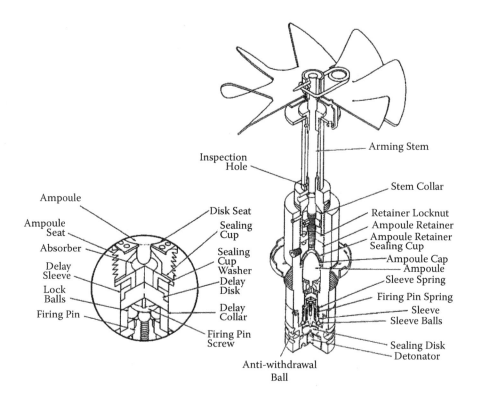

Figure 3.21 M123 chemical long-delay fuze. Note the "anti-withdrawal ball" near the base. (From U.S. military TM.)

largely obsolete. However, they will be encountered well into the future and pose significant hazards, thus requiring introduction. To prevent successful EOD action, most designs include an anti-withdrawal feature that will immediately function the fuze if removal is attempted (Figure 3.21).

General identification features associated with chemical long-delay fuzes include:

- **Construction:** Multipiece.
- **Materials:** Steel body; there may be a copper component where the ampoule sits.
- **Markings:** Nomenclature may be stamped on observable sections of the fuze.
- **Other:** The time delay is set prior to fuze installation. Oftentimes these fuzes are designed to look similar to other fuzes; positive identification is crucial.
- **Common employment:** Bombs.
- **General safety precautions** for chemical long-delay fuzes include:
- HE, frag, movement, C/S, and W/T.

3. Proximity or variable time (VT) Fuzes: Are capable of sensing a target and their proximity to it. They are designed to function the munition during flight at a specific distance from the target. The intent is an airburst at a distance that maximizes the munition's effect on the target. There are two basic configurations, both of which are categorized as VT fuzes:

1. **Active VT fuzes** contain electronics packages that produce an active signal and receive the signal back after reflecting off the target. When the signal reflection matches the sensor setting, a circuit completes and functions the fuze.
2. **Passive proximity fuzes** contain electronics packages that use signals from the target such as heat and sound. When incoming signals match the sensor setting, a circuit completes and functions the fuze.

Proximity or VT fuzes are powered by wet cell, dry cell, and thermal batteries, or wind-driven generators. There are many different designs incorporating a variety of mechanical and/or electrical arming actions or sequences, including pins or wires that must be removed, setback or centrifugal force requirements, electrical power, or arming vanes that must be rotated. Some VT fuzes have graduated time rings similar to a mechanical time fuze and employ a timed self-destruct backup feature. There are also multi-option VT fuzes that can be set for impact as a backup feature.

Figure 3.22 VT fuze—older design incorporating a wet cell power source. (Author's photograph.)

General identification features associated with proximity or VT fuzes include (Figures 3.22–3.23):

- **Construction:** There are three configurations:
  - A single-element, multi-piece fuze in the nose or front of the ordnance that contains all required fuzing components.
  - A two-element, multi-piece fuze with a nose- or front-mounted proximity-sensing element electrically connected to a base or tail element containing the remaining fuze components (see Figure 3.1).
  - Side-mounted configurations are common on high-speed anti-aircraft missiles.

Figure 3.23 U.S. proximity or VT fuzes; from left: M532 projectile fuze, M429 rocket fuze, and M414A1 rocket fuze. (Author's photograph.)

- **Materials:** Older designs were made of steel with a loop-antenna shape (Figure 3.24). This design was replaced with a plastic cap or other material cap/shield capable of allowing a signal to pass through it while protecting the electronic components. These covers can be opaque, translucent, or a number of colors. The lower body of a nose- or front-mounted element may have a steel or aluminum lower body.
- **Markings:** Nomenclature may be stamped on observable sections of the fuze. On two-element configurations, the base or tail element may not have any markings.
- **Other:** The distance in which the fuze will function from the target is preset.
- **Common employment:** Bombs, projectiles, rockets, and missiles.
- **General safety precautions** for proximity or VT fuzes include:
- HE, frag, movement, VT, EMR, static, and W/T

Note: There are VT fuzes with impact backups containing a C/S hazard.

4. Pressure fuzes are usually designed to function a munition when it comes in contact with the target. Most commonly associated with landmines, boobytraps, and underwater ordnance, pressure fuzes can be located on the front, rear, top, bottom, side, or inside of a munition. There are many different mechanical and

Figure 3.24 British no. 952 MK1, VT fuze—older, yet durable design that is still used today. (Author's photograph.)

electronic designs incorporating a variety of arming actions or sequences, but most involve pins, wires, or clips that must be removed. There are also electronic pressure fuzes capable of being programmed to arm or self-destruct, which will be covered in later chapters. Six fuze types will be covered under the pressure group. Due to the straightforward nature of pressure fuzes, each type will be briefly covered; however, when ordnance employing pressure fuzes is covered in later chapters, additional information relevant to the specific fuze types will be provided.

   4a. Direct pressure fuzes are designed to function when the pressure required to overcome a positive block or spring is applied. Coiled and Belleville springs are commonly used for this application and direct pressure fuzing is the most common fuze employed in landmines.

   General identification features associated with direct pressure fuzes include:

- **Construction:** Multipiece with a "push-button" configuration, or prongs which are common on some mines.
- **Materials:** Metal or plastic is most common.
- **Markings:** Nomenclature may be stamped on observable sections of the fuze.
- **Common employment:** Landmines and underwater ordnance.

- **General safety precautions** for direct pressure fuzes include:
  - HE, frag, movement.
  - EMR; static for fuzes with electrical components.

4b. Pressure or tension release fuzes are designed to function when pressure is taken away or released. The secondary fuze wells in anti-tank (AT) mines are designed for this employment.

General identification features associated with pressure or tension release fuzes include:

- **Construction:** Multi-piece configuration.
- **Materials:** Metal or plastic is most common.
- **Markings:** Nomenclature may be stamped on observable sections.
- **Other:** Presence of prongs or a wire running to an unknown item.
- **Common employment:** Landmines.
- **General safety precautions** for pressure or tension release fuzes:
  - HE, frag, movement.
  - EMR, static for fuzes with electrical components.

4c. Hydraulic pressure fuzes: Are designed to function when the pressure required to overcome a positive block or spring is applied, but there is an additional time requirement. The "time" component allows for target specificity; for example, wheeled vehicles can be ignored while tracked vehicles can still be targeted as they maintain pressure for a longer period. This design also offers a fuze the ability to count targets and function after a specific number.

General identification features associated with hydraulic pressure fuzes include (Figure 3.25):

- **Construction:** Multi-piece configuration.
- **Materials:** Metal, plastic, or rubber.
- **Markings:** Nomenclature may be stamped on observable sections.
- **Other:** Tubes or hose-like components may be present.
- **Common employment:** Landmines.
- **General safety precautions** for hydraulic pressure fuzes include:

Figure 3.25 British L9 bar mine incorporating an L89 or L90 hydraulic pressure fuze. (Author's photograph.)

- HE, frag, movement.
- EMR, static for fuzes with electrical components.

4d. Barometric pressure fuzes detect pressure changes associated with altitude and are designed to function at a specific height as a bomb descends. These fuzes are uncommon.

General identification features associated with barometric pressure fuzes include (Figure 3.26):

- **Construction:** Multi-piece configuration.
- **Materials:** Metal.
- **Markings:** Nomenclature may be stamped on observable sections.
- **Common employment:** Bombs.
- **General safety precautions** for barometric pressure fuzes include:
  - HE, frag, movement.

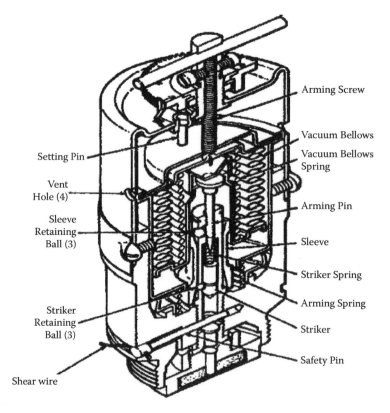

Figure 3.26 Internal configuration of a barometric pressure fuze. (From U.S. military TM.)

4e. Cumulative pressure fuzes are designed to function when the pressure required to overcome a positive block or spring is applied, but they have the ability to account for previously applied pressures. For example, if 10 lb of pressure is applied to a direct pressure fuze requiring 30 lb of pressure, it will not function, even if 10 lb is applied multiple times. However, if 10 lb of pressure is applied three times to a cumulative pressure fuze requiring 30 lb of pressure, it will function.

General identification features associated with cumulative pressure fuzes:

- **Construction:** Multi-piece configuration.
- **Materials:** Metal or plastic.
- **Markings:** Nomenclature may be stamped on observable sections.
- **Common employment:** Submunitions.
- **General safety precautions** for cumulative pressure fuzes include:
- HE, frag, movement.

4f. Hydrostatic pressure fuzes are designed to sense pressure changes as a munition sinks in water and function at a specific depth.

General identification features associated with hydrostatic pressure fuzes include:

- **Construction:** Multi-piece configuration.
- **Materials:** Steel, brass.
- **Markings:** Nomenclature may be stamped on observable sections.
- **Common employment:** Depth charges, depth bombs, and signal devices; they are also employed with other underwater fuzing systems.
- **General safety precautions** for hydrostatic pressure fuzes include:
  - HE, frag, movement.
  - EMR, static for fuzes with electrical components.

5. Influence fuzes are designed to function a munition when specific magnetic, acoustic, or seismic thresholds are presented. Influence fuzing systems can be set or programmed for specific targets, are very sophisticated, and exemplify the precision aspect of fuze functioning. There are many different designs that may be located in the front, rear, side, inside or in some cases, have an external component tethered to the munition. Arming actions or sequences can include pins or wires that must be removed; retardation, setback, or centrifugal force; electrical power; arming vanes that must be rotated; or impact with the ground or water. Self-destruct (S/D) features are commonly employed as a backup to ensure the munition functions on or near a target. In other cases, a S/D is used to stop other countries from exploiting the fuze.

It is not uncommon for a combination of influence types to be employed in a single fuze. For example, a seismically armed and magnetically fired fuze may be able to "sleep," thus saving battery power until a tank or other target with a specific seismic signature "wakes it up," powers, and arms the magnetic influence feature. The capability to sleep and conserve power in this manner may allow these fuzes to provide a threat for decades.

Most often employed with underwater ordnance, landmines, and submunitions, these fuzes are also used to turn a conventional bomb into a magnetic influence land or sea mine. These fuzes are always considered to be powered and actively seeking a target until proven otherwise. Three fuze types will be covered under the influence group.

5a. Magnetic influence fuzes are designed to detect changes associated with the amount of ferrous metal in their vicinity and function when those changes match the programmed setting. General identification features associated with magnetic influence fuzes include:

- **Construction:** Multi-piece including an arming section, booster, and magnetic sensor.
- **Materials:** Metal and plastic.
- **Markings:** Nomenclature may be stamped on observable sections.
- **Other:** Positive identification from a distance is crucial.
- **Common employment:** Bombs, landmines, and sea mines.
- **General safety precautions** for magnetic influence fuzes includes:
- HE, frag, movement, EMR, static, influence, and W/T.

5b. Acoustic influence fuzes are designed to detect changes associated with the ambient sounds in its vicinity and function when those changes match the programmed setting.

General identification features associated with acoustic influence fuzes include:

- **Construction:** Multi-piece, including an arming section, booster, and sensor or hydrophone to detect sounds.
- **Materials:** Metal and plastic.
- **Markings:** Nomenclature may be stamped on observable sections.
- **Other:** Positive identification from a distance is crucial.
- **Common employment:** Sea mines and landmines.
- **General safety precautions** for acoustic influence fuzes include:
  - HE, frag, movement, EMR, static, influence, and W/T.

5c. Seismic influence fuzes are designed to detect changes associated with vibrations caused by low-frequency sound or movement in the vicinity and function when those changes match the programmed setting. This allows the magnetic fuze to "sleep" and conserve power until a potential target is in the vicinity.

General identification features associated with seismic influence fuzes include:

- **Construction:** Multi-piece, including an arming section, booster, and sensor or geophone to detect vibrations.
- **Materials:** Metal and plastic.
- **Markings:** Nomenclature may be stamped on observable sections.
- **Other:** Positive identification from a distance is crucial.
- **Common employment:** Sea mines and landmines.
- **General safety precautions** for seismic influence fuzes include:
  - HE, frag, movement, EMR, static, influence, and W/T.

6. Anti-disturbance (A/D) fuzes are considered a boobytrap designed to disrupt or kill those attempting to move the munition. They can be used as the primary or secondary means of functioning; there are electrical and mechanical versions that arm after deployment, and some are disguised to appear as a conventionally fuzed munition. Depending on the design, A/D fuzes can be located in the front, rear, side, or inside of a munition and incorporate a variety of mechanical and/or electrical arming actions or sequences, including pins or wires that must be removed, retardation, setback or centrifugal force requirements, electrical power, or arming vanes that must be rotated. As with all fuzes, positive identification is required to ensure the appropriate safety precautions are adhered to.

General identification features associated with A/D fuzes include:

- **Construction:** Usually internal to a munition or in a section of the fuze that cannot be seen.
- **Materials:** Metal and plastic.
- **Markings:** Nomenclature or markings are seldom present.
- **Other:** Positive identification of a munition that may contain an A/D fuze is crucial.
- **Common employment** is with landmines, submunitions, and underwater ordnance.
- **General safety precautions** for A/D fuzes include:
  - HE, frag, movement, EMR, static, and W/T.

Note: Unlike an anti-withdrawal feature that requires a fuze to be unscrewed or withdrawn, an A/D fuze will function with the slightest movement.

Note: Commonly used in conjunction with a self-destruct feature.

Closing

The accurate identification of a munition provides tremendous insight on the fuzing possibilities. However, fuzing systems are constantly updated and pose a somewhat unknown threat aspect to all munitions.

Ordnance Category—Projectiles

4

No one accuses the gunner of maudlin affection for anything except his beasts and his weapons. He serves at least three jealous gods—his horse and all its saddlery and harness; his gun, whose least detail of efficiency is more important than men's lives; and, when these have been attended to, the never-ending mystery of his art commands him.

Rudyard Kipling

Introduction

Interestingly, Kipling used the words "mystery" and "art" when describing an artilleryman. The mystery may have been the manner and variety in which the ordnance functioned and the art may have alluded to the accuracy in which projectiles could be deployed at great distance. When considering that artillery is referred to as the "king of battle" and is believed to have accounted for more combat deaths than any other ordnance type, Kipling's choice of words is insightful.

Basically described, a projectile is a munition projected by external force and continuing in motion by its own inertia. This description includes stones thrown from a medieval trebuchet and the first black powder-filled exploding cannonballs to the many different ordnance categories of today. For this chapter, the defining factors that categorize a munition as a "projectile" are that

1. The body being projected is fired down a barrel or tube by gas pressure generated from a propellant charge.
2. The propellant charge is the munition's primary means of deployment.
3. The body being projected does not have an attached motor as a primary means of propulsion.

There are exceptions to this definition, such as projected grenades that meet all defining characteristics of a projectile, but are categorized under grenades, and Rocket-Assisted Projectile (RAP). A RAP has a rocket motor designed as a secondary means of propulsion to increase the range of a projectile after it has been fired (Figure 4.1). The most common munitions

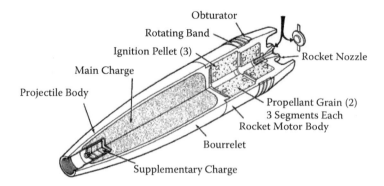

Figure 4.1 Line drawing of an M549, 155mm RAP configuration. (From U.S. military TM.)

meeting the three defining factors and categorized as projectiles are fired by mortars, howitzers, recoilless rifles, and guns.

As projectiles were the original ordnance items, their designs have influenced the shapes and naming conventions of all ordnance categories. In this chapter, many aspects of construction will be covered that will be seen again in the chapters on rockets and missiles.

Delivery Systems

There are four common delivery systems used to deploy projectiles, all of which provide design features that will help identify a munition.

Mortars: Are designed to launch a projectile over a barrier such as a wall or terrain feature and tend to fire a slower moving projectile in a high, arcing trajectory. Some mortar tubes are rifled, but most have a smooth bore.

Howitzers: Constitute the majority of field artillery pieces and fire a medium velocity projectile at a much lower trajectory than a mortar. Howitzers normally have rifled bores.

Recoilless rifles: Fire projectiles at a high velocity with a flat trajectory. As the name implies, these weapons have no recoil as the energy is blown out of the opposite end of the weapon from which the projectile exits. Recoilless rifles normally have rifled bores, but there are smooth-bore versions classified as recoilless guns.

Guns: Fire projectiles at a high velocity with a flat trajectory. They are commonly fired from ships, tanks, and field guns that may have a smooth bore to decrease friction and increase velocity. Guns can have rifled or smooth bores.

Projectile Configurations

Due to the diverse number of projectile designs, a configuration classification is provided to assist with this ordnance category. There are four basic configurations associated with projectiles (Figure 4.2):

Fixed: A projectile is crimped and secured to a cartridge case. The propellant charge cannot be accessed or altered. The munition is loaded in the same manner as a bullet. A crimping ring or rings (Figure 4.2) are common on fixed projectiles.

Separated: A projectile is shipped separately from the cartridge case. The open end of the cartridge case is sealed and the propellant charge cannot be accessed or altered. To load, the projectile is loaded and the cartridge containing the propellant is loaded behind it.

Semifixed: A projectile is set into the cartridge case, but can be removed to allow access to the propellant charge for adjustments prior to loading. To load, the projectile is set into the cartridge case after the propellant charge has been adjusted and the complete munition is loaded.

Separate loading: A projectile is separate from the propellant charge and there is no cartridge case. Loading involves the projectile being loaded and the bags containing the propellant charge being placed behind it, which allows the quantity of propellant to be adjusted easily.

Key Identification Features

A projectile is a teardrop- or square-shaped munition with fins for stabilization, or a finless munition that employs a rotating band for spin stabilization. Early cannonball type projectiles were essentially thrown toward a target without a means of stabilization. Lacking stabilization, the accuracy and range of these munitions was poor. Development of cylindrically shaped projectiles with fins for stabilization and rifled bores capable of imparting spin to gyroscopically stabilize flight greatly enhanced the range, accuracy, and velocity of projectiles. These modern advancements also provide identifying characteristics that greatly assist in determining the group and associated safety precautions.

Worldwide there have been tens of thousands of different projectile designs over the years, which may make positive identification impossible, ranging in size from 20mm to the behemoth 600mm German mortar "Thor." Determining the group and safety precautions associated with an unknown projectile may be possible due to its construction charactaristics (Figure 4.3).

A - Fuze
B - Booster
C - Projectile
D - Ogive
E - Bourrelet
F - Bursting Charge
G - Rotating Band
H - Crimp
J - Base Cover
K - Cartridge Case
L - Propelling Charge
M - Primer
N - Lifting Plug
P - Grommet
Q - Igniter
R - Cased Propelling Charge
S - Closing Plug
T - Distance Wad
U - Igniter Charge Assembly

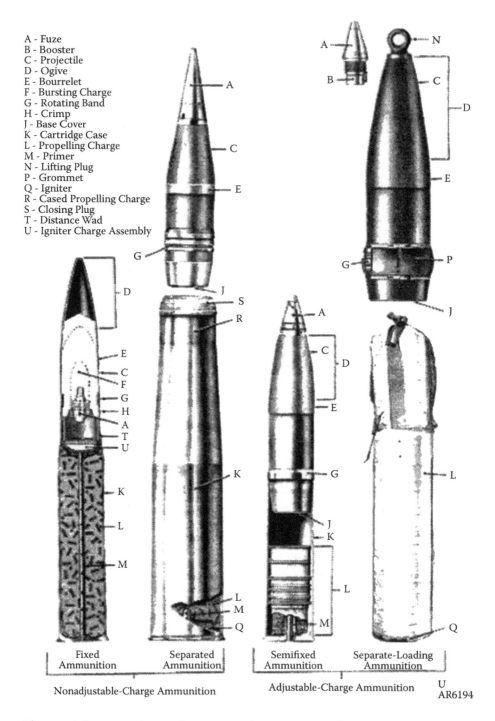

| Fixed Ammunition | Separated Ammunition | Semifixed Ammunition | Separate-Loading Ammunition |

Nonadjustable-Charge Ammunition   Adjustable-Charge Ammunition

AR6194

Figure 4.2 Projectile configurations. (From U.S. military training manual [TM].)

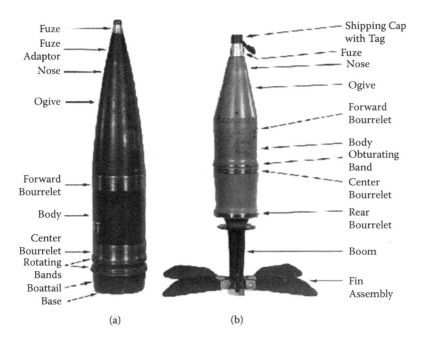

Figure 4.3 Basic configuration of a spin stabilized projectile (left) and a full or partially fin stabilized projectile (right). (From U.S. military TM.)

Projectile Sections

Ogive is the section forward of the bourrelet, and if present it is below the fuze or fuze adapter. The ogive may be flat, rounded, conical, or spiked. The length and shape of the ogive greatly influence the flight characteristics of a projectile. Most nose-fuzed projectiles have a conical-shaped ogive that continues the contour of the nose fuze. Flat ogives are common on canister type anti-personnel (APERS) projectiles that travel very short distances. Spikes, hammer rings, adapters, and other configurations provide insight into the group to which a munition belongs (Figures 4.4–4.9).

Bourrelet is the section between the ogive and the body of a projectile. Some projectiles have a second bourrelet on the body, just forward of the rotating band or just below the rotating band on the base. The bourrelet is a slightly raised surface on the projectile that contacts the inside of the bore to act as a guide when the projectile is fired. When measuring the diameter of a projectile, take the measurement at the bourrelet as this is the true diameter. If a bourrelet is not present, measure at the junction of the ogive and the body (Figures 4.10–4.12).

Warhead or **body** is oftentimes used interchangeably; for projectiles, this is the cylindrical section of the projectile between the forward bourrelet and

Figure 4.4 Hammer rings. Key identification features of an Armor Piercing (AP) munition. (Author's photograph.)

Figure 4.5 A smooth, rounded ogive of light material is consistent with a High Explosive Plastic (HEP) projectile. (Author's photograph.)

Figure 4.6 A standoff spike design is a key identification feature for a High Explosive Anti-Tank (HEAT) munition. (Photograph courtesy of Didzis Jurcins.)

Figure 4.7 A flat ogive of thin metal construction indicates a canister or shot munition that functions similarly to a very large shotgun. (Author's photograph.)

Figure 4.8 An elongated ogive with a constant angle and no breaks is consistent with a High Explosive (HE) munition. (Author's photograph.)

Figure 4.9 An elongated ogive with a constant angle and an adapter can mean a few things. An adapter may allow for the employment of a variety of fuzes for an HE munition, or it may be an adapter booster used to seal in a White Phosphorus (WP) or chemical filler. Key identification features are a break where two components meet, and spanner holes or wrench slots. (Author's photograph.)

Figure 4.10 Measure the true diameter of a munition at the bourrelet. Adjust external calipers until them fit snugly on the bourrelet. (Author's photograph.)

the rotating band. Depending on design, the body may be slightly smaller in diameter than the bourrelet or taper down in diameter toward the base. By being smaller than the bourrelet, the body does not contact the bore, which greatly reduces wear on the barrel. The term warhead is often applied when a projectile contains fins and a rocket-assist motor, but still alludes to the

Figure 4.11 Move the calipers to a metric ruler to ascertain the diameter. (Author's photograph.)

Figure 4.12 Close up of measurement, 100mm. Russian OF.412 projectile. (Author's photograph.)

section containing energetic materials, chemical or inert filler. One of the two most common areas to find stamped markings is on the body. Different countries favor various locations, but common areas are just above or below the rotating band(s) and the midbody (Figure 2.4).

Rotating band, gas-check band, and **obturator ring:** Depending on the delivery system, there are a few different band designs that serve somewhat similar purposes. All rotating bands, gas-check bands, and obturator rings are designed to trap the gas generated from the propellant behind the projectile as it moves down the bore to maximize the range of the projectile. Rotating bands, which are usually made of copper (older designs used lead), are also designed to engage the riflings in a bore and impart spin as a projectile moves down the bore, which will provide stabilization during flight. Rotating bands may be narrow or wide, smooth, or have circumferential grooves called "cannelures" machined into them to catch shavings and reduce friction when fired. New or specifically designed projectiles may have rotating bands made from sintered iron or plastic (Figures 4.13–4.19).

Mortars usually have gas-check bands or an obturator ring versus a bourrelet (Figures 4.17 and 4.18). But there are spin stabilized mortars with rotating bands (Figure 4.19). Howitzer projectiles usually employ a single

Figure 4.13 A lead rotating band is consistent with ordnance manufactured prior to 1900, as seen on this British Armstrong 3 in. projectile. (Author's photograph.)

Figure 4.14 Rotating band of an unfired artillery projectile. Note the smooth appearance of the band. (Author's photograph.)

Figure 4.15 Rotating band of a fired artillery projectile. Circled in yellow is the scored and distorted bottom edge of the band. Compare with Figure 4.20. (Author's photograph.)

Figure 4.16 Rotating band with two cannelures on an unfired artillery projectile. Additionally, the two indents or grooves below the rotating band are "crimping rings," which are commonly associated with fixed projectiles. (Author's photograph.)

Figure 4.17 The obturator ring is the white plastic ring toward the center of the projectile. (Author's photograph.)

rotating band approximately one inch in width; however, multiple thin bands or a single, very wide band is also common. If a projectile is intact and the rotating band is missing, its "seat" or the groove in the body where it would have sat will be observable. In this situation, the pattern of indentations in this groove will be very helpful in the identification process.

Recoilless rifle projectiles may have a single rotating or a very wide bourrelet. Many rotating bands on recoilless rifle projectiles have a unique characteristic in that they are prescored in order to be fitted into the riflings of a bore (Figure 4.20).

Gun projectiles are commonly fired from ships, tanks, and other field pieces. In order to increase velocity, these projectiles usually have multiple, large, and prominent rotating bands. Smooth-bore guns are often used to decrease friction and increase velocity, but projectiles from these guns require fins for stabilization as they do not have rotating bands.

Figure 4.18 Gas-check bands on a Russian 82mm mortar. (Author's photograph.)

Figure 4.19 120mm spin stabilized illumination mortar, note pre-scored rotating band and unique umbrella-like symbol used to identify an illumination munition. (Author's photograph.)

Figure 4.20 The pre-scored rotating band on a recoilless rifle projectile. Circled in yellow is the undistorted bottom edge of the scored band. Compare with Figure 4.15. Note the crimping ring below the rotating band. (Author's photograph.)

Base is the bottom, rear, or end of a projectile that may or may not contain a fuze. There are a number of key identification features associated with the base of a projectile that can assist in the identification process. Projectiles may have a square, boat-tailed, recessed, or curved base that can be welded, cast, pressed with a lip, flush, countersunk, or have a thin metallic cover. The presence of crimping rings between the rotating band(s) and base of the projectile may allow differentiation between similar projectiles (Figures 4.21). Other features include fins, a tracer element, venturi, or nozzles that obscure the base, thus hindering a determination of the presence of a base fuze (Figures 4.22–4.26).

Figure 4.21 Three different tracer element designs on 85mm APHE projectiles, all three are mounted over a base fuze. Note the crimping rings and shape of the base below the rotating bands. (Author's photograph.)

Figure 4.22 "Fin and Spin" stabilized 90mm Yugoslavian HEAT, M74 projectile. Note the rotating band used to impart spin, and fins which open outward after leaving the bore. (Author's photograph.)

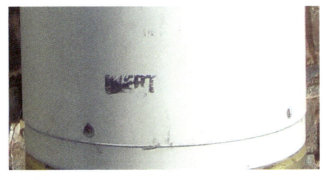

Figure 4.23 Square base with pressed and pinned base plate. Note holes in the side for pins. (Author's photograph.)

Figure 4.24 Tapered, solid base. (Author's photograph.)

Figure 4.25 Solid base plate. (Author's photograph.)

Figure 4.26 Counter-sunk, threaded base plate. Note spanner holes. (Author's photograph.)

Figure 4.27 A 155mm South African M1A1 HE projectile, with a lifting ring in the fuze well. Note the brown band and break just below the rotating band indicating the presence of propellant; in this case, a base-burner element. The "flutes" on the body are a unique identification feature for this projectile. (Author's photograph.)

Conversely, determining if a nozzle represents the presence of a tracer, RAP, or base burner element can be difficult. Tracer elements vary in size and shape and may be part of a base fuze. The venturi or nozzle of a rocket-assisted projectile such as the M549 in Figure 4.1 can be identified by inspecting the body for a major break on the lower half of the body. Containing 7 lb of solid rocket propellant, which is a separate component from the forward portion of the projectile, the two-component configuration of the M549 provides an example of an identifiable seam or break in the body of the projectile. Additionally, RAPs use rocket propulsion as a means of increasing range; as such the nozzle will have a tapered appearance consistent with a rocket venturi as it is designed to produce thrust. Base burner or fumer elements also employ propellant, but in a very different way. As a projectile moves through the air, a vacuum develops drag on the base, accounting for 50% of the overall drag associated with the trajectory of a projectile. During flight, a base burner bleeds hot gas into the area where the vacuum forms, breaking up the drag and thus greatly increasing the range of the projectile (Figure 4.27).

Fin assemblies: Many projectiles employ fins or a combination of spin and fins for stabilization. Projectiles employing a spin–fin combination normally have rotating band(s) located near the base. Fins offer many identifiable features; for example, if a deployed mortar is embedded in the ground and only the fins can be seen, determining the diameter of the projectile is possible by measuring the diameter of the fins, which are the same diameter as the body. This is also true for many recoilless rifle projectiles. A key ID feature to differentiate a projectile with fins from a rifle grenade is the ignition holes in the tail boom (Figures 4.28–4.30), a feature not present on rifle grenades.

Fuze: Many projectile fuzes require three actions to arm that coincide with the way the munition is deployed—for example:

- A **spin**-stabilized projectile fuze usually requires three actions: (1) **setback** from the acceleration of firing, (2) **centrifugal force** from spin, and (3) **time of flight** during which the actions of 1 and 2 release locks or clockwork mechanisms, initiate pyrotechnic delays, and other actions to complete the arming sequence.

Figure 4.28 Fin stabilized mortars. Left: An M67, 82mm Illumination mortar with propellant increments. Right: An 81mm high explosive mortar with exposed ignition holes.

Note: The fins are the same diameter as the projectiles. (Author's photograph.)

- A **fin**-stabilized projectile that does not spin may employ a fuze with (1) **pins** or **clips** that must be removed prior to firing, and then (2) **setback** from the acceleration of firing, and (3) **time of flight** during which the actions of 1 and 2 release locks or clockwork mechanisms, initiate pyrotechnic delays, and other actions to complete the arming sequence.

Projectile nose fuzes are usually observable, but impact with a hard surface may render identification difficult. If a nose fuze is sheared off flush

Figure 4.29 82mm Chinese HEAT, Type 64, spin stabilized recoilless rifle projectiles. Left: With propellant in place. Right: With ignition holes exposed. Note: The pre-scored rotating band at base encircling the fins are the same diameter as the projectile. (Author's photograph.)

with the fuze well, the components required to function the fuze may still be present. If a nose or base fuze can be seen, the wrench flats, spanner holes or slots, and the overall construction will provide relevant information to its identity. Additionally, the presence of a fuze adapter or booster adapter between the fuze and the projectile may help identify both the fuze and the projectile.

However, there are projectile fuzes that are internal to the munition and offer few clues to their existence. An example would be the Base-Detonating (BD) element of a Point-Initiating Base-Detonating (PIBD) fuze. Always consider the possibility of hidden or unseen fuzing.

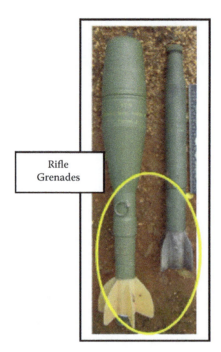

Figure 4.30 Two Yugoslavian Rifle Grenades are displayed to illustrate a key identification feature. Note the lack of ignition holes on or above the fins and that the fins are not the same diameter as the widest section of the body. (Author's photograph.)

The Seven-Step Practical Process Applied to Projectiles

Examples of different designs, features, color codes, markings, and construction features are provided throughout this chapter.

Step 1: Approach and initial interrogation. Attempt to identify a munition at a distance with the use of binoculars. If an approach is made, avoid all venturis and fuze-sensing elements. Armed and active or damaged sensing elements may "see" a person approaching, consider the person a valid target, and function as designed.

At a minimum, measurements must be taken of the major diameter at the bourrelet and the overall length of the body from the base to the fuze well. If possible, measure the location and width of the rotating bands, gas-check bands, obturator ring, and crimping rings. If there is more than one, measure the distance between rotating bands and crimping rings; if present, measure and count the number of gas-check bands.

Look for stamped data on the munition and, if present, focus on the area just above and below the rotating band. All findings, including measurements, color codes, markings, key identifying features, and any possible damage, are documented and the munition is photographed.

In addition to the overall configuration, there are three features that will greatly assist in answering steps 2, 3, 5, and 7:

1. The diameter
2. The overall length
3. The method of stabilization (i.e., spin, fin, or a combination of both)

Step 2: Determine fuze type and condition. If a projectile has been deployed, the fuze is considered to be armed (step 5). If a fuze is damaged, pins have been removed, or any alterations have been made to the munition, it is considered armed. If visible, measurements for the fuze are taken separately from the munition.

Step 3: Determine ordnance category. This category covers all ordnance that is fired down a barrel or tube by gas pressure generated from a propellant charge as its primary means of deployment. Some projectiles are fitted with rocket motors, but these are not the primary means of propulsion.

Step 4: Determine ordnance group. Identifying characteristics associated with each projectile group will be covered throughout this chapter, many of which are consistent with the grouping characteristics of other categories.

Step 5: Determine if the munition was deployed. Inspect the projectile for impact-related damage; soot on the base, fins, or tail boom from propellant charge; missing pins or clips; and, if present, scoring on the rotating band.

Note: The rotating band of a recoilless rifle projectile is easily mistaken for a scored rotating band. Figures 4.14 and 4.15 provide an example of an artillery rotating band unfired and after being fired. Figures 4.20 and 4.29 provide examples of a recoilless rifle rotating band, which will appear the same both prior to and after being fired.

Step 6: Determine safety precautions that apply to the munition. The safety precautions for the projectile groups are covered in this chapter. Chapter 3 addresses the safety precautions associated with various fuzes.

Note: Adhere to all safety precautions that apply.

Step 7: Identify the munition. Apply the totality of all construction characteristics and other identifying features to determine the group to which a projectile belongs and, if feasible, positively identify the munition and all possible fuzing configurations.

Groups

Per the definition provided in the Introduction of this chapter, the projectile category encompasses thousands of different ordnance items. In order to provide a coherent flow, the category projectile is divided into the following primary and supplemental groups:

1. High Explosive (HE).
   a. HE/fragmentation (frag).
   b. High-Explosive Incendiary (HEI).
   c. High-Explosive Plastic (HEP).
   d. HE-RAP.

2. High-Explosive Anti-Tank (HEAT).
3. Guided projectiles.
4. Armor Piercing (AP).
   a. AP.
   b. APHE.

5. Anti-personnel (APERS).
6. Dispenser and Improved Conventional Munition (ICM).
7. Smoke.
   a. Bursting smoke.
   b. Burning smoke.
   c. Riot control.

8. Illumination.
9. Practice.
   a. With and without spotting charges.
   b. Drill and dummy.

Groups

1. Projectile, High Explosive (HE): All four of the projectile configurations covered under this general group are designed to explode and produce destructive effects through blast pressures and fragmentation. Explosive fillers in these projectiles range from less than an ounce to hundreds of pounds and may be solid, pliable, or in liquid form. Design features provide evidence to assist in the identification process.

1a. **Projectile, HE/frag** represents the most straightforward design of an explosive munition, a projectile body, single or dual fuzed, and high-explosive filler.

Figure 4.31 Often confused for an RPG HEAT rocket, this 73mm Russian PG-9 does not have a motor in the tail boom and is fired as a projectile by the propellant charge attached to its base. Though the warhead shape is consistent with a HEAT design, this projectile is an HE munition and lacks the tell-tale break in the major diameter, consistent with a HEAT munition. (Author's photograph.)

General identification features associated with HE/frag projectiles include (Figures 4.31 and 4.32):

- **Appearance and materials:**
  - Solid, one-piece body of robust construction with rotating band(s), gas-check bands or an obturator ring.
  - The United States does not commonly use fuze adapters with HE ordnance, but other countries do. When present, top-down

Figure 4.32 Assortment of mortars, from left to right: U.S. 60mm HE, Israeli 60mm HE, U.S. 60mm Practice, Chinese 50mm HE, and Japanese 2 inch HE. Note: "Ni," pronounced "knee" is the number 2 in Japanese, resulting in a common reference to a "Ni mortar." (Author's photograph.)

slots or side spanner holes on a fuze adapter may indicate if the projectile is filled with HE or WP.

- A welded base on a U.S. projectile is positive identification for HE. However, many countries use a solid base design (Figures 4.24 and 4.25).
- Fin assemblies for recoilless rifle projectiles have ignition holes in the tail boom assembly. Fin assemblies for mortars have ignition holes in the tail boom assembly or between the fins (Figures 4.29 and 4.32).
- If ignition holes are present in the tail boom or fins, a primer and ignition assembly should be located in the base of the tail boom.
- If a tail boom or fins are present, there will not be a venturi consistent with a rocket or an open tube consistent with a rifle grenade at the base.
- Naval projectiles tend to have a heavier body construction and extremely wide or multiple rotating bands to withstand the higher firing velocities.
- Most HE mortars have a teardrop shape. Exceptions include the U.S. 4.2 in. (107mm) and 120mm mortars (Figure 4.19).

- **Markings:** A green or black body with yellow markings or a gray body with black markings is common. A brown band (Figure 4.27) may be present to designate a base burner or rocket motor. Other colors, stamped or stenciled markings, and symbols may also be present.
- **Common fuze configurations:** Point-Detonating (PD), BD, and Variable-Time (VT) fuzing.
- **General safety precautions** for HE/frag projectiles include:
  - HE, frag, movement.
  - Safety precautions for the fuze if present.

1b. Projectile, High-Explosive Incendiary (HEI): Contains an HE main charge enhanced with incendiary materials. Common incendiary materials include aluminum, magnesium, and zirconium, which can be blended with the main charge explosive prior to filling the projectile. Other means of enhancing the main charge include adding incendiary pellets or a pyrophoric liner elsewhere on the warhead.

General identification features associated with HEI projectiles include:

- **Appearance and materials:** Construction and identification features are consistent with HE projectiles.
- **Markings:** A green or yellow body with yellow and red markings is common. Other colors such as an all-black body (Figure 4.33), stamped or stenciled markings, and symbols may also be present.

Figure 4.33 Left: Russian 23mm HEI. Right: U.S. 25mm HEI designated with red markings. (Author's photograph.)

Red is usually used to designate an incendiary projectile, but some countries use red to identify HE or HEAT munitions, and other countries add incendiary materials without an identifying color.

- **Common fuze configurations:** PD, BD, and VT fuzing.
- **General safety precautions** for HEI projectiles include:
  - HE, frag, movement, fire.
  - Safety precautions for the fuze if present.

1c. Projectile, High-Explosive Plastic (HEP): Contains an HE main charge in a configuration designed to produce a specific effect. Also known as a "squashhead," a HEP projectile has a thin metal ogive and body filled with a pliable explosive and a solid base, which houses a BD fuze (Figure 4.34). Upon impact with an armored target, the forward end of the projectile is "squashed" in a similar configuration to that of a wet paper towel thrown against a wall. The BD fuze functions, detonating the main charge and producing a spalling

Figure 4.34 Cutaway of a 106mm U.S. M346A HEP Projectile. (Author's photograph.)

effect on the inside of the armored vehicle. Against modern armor, HEP projectiles are ineffective and generally an obsolete design.

General identification features associated with HEP projectiles include (Figure 4.34):

- **Appearance and materials:**
  - Heavy-duty construction.
  - Short, dome-shaped ogive.
  - Single-piece body construction.

- **Markings:** A green body with yellow markings and black markings is common. Other colors, stamped or stenciled markings, and symbols may also be present.
- **Common fuze configurations:** BD fuzing.
- **General safety precautions** for HEP projectiles include:
  - HE, frag, movement.
  - Safety precautions for the fuze if present.

Note: There is a 105mm WP projectile with the construction characteristics of a HEP projectile—reinforcing the importance of positive identification.

1d. Projectile, Rocket Assisted (RAP): Contains a rocket motor in the base that is initiated when the projectile is fired, thus greatly increasing its range (Figure 4.1). When used with a high-explosive projectile, the naming designation used is High-Explosive Rocket Assisted (HERA).

General identification features associated with RAP projectiles include:

- **Appearance and materials:**
  - Heavy-duty construction.
  - Nose fuze only.
  - One or possibly two breaks in the bottom half of the body.
  - Venturi on the base.

- **Markings:** A green body with yellow markings is common. There may be brown markings below the rotating band for the motor. Other colors, stamped or stenciled markings and symbols may also be present.
- **Common fuze configurations:** PD, VT fuzing.
- **General safety precautions** for RAP projectiles include:
  - HE, frag, movement, and ejection.
  - Safety precautions for the fuze if present.

2. Projectile, High-Explosive Anti-Tank (HEAT) contains a shaped charge warhead and is designed to defeat armored and other hardened targets. In order to allow the shaped charge to maximize performance and penetration, a standoff spike or hollow area is located in front of the cone. Due to the space needed to provide a standoff, HEAT projectiles have lower explosive weights than HE munitions of similar size. Most HEAT projectiles are between 40mm and 155mm in diameter. The majority of HEAT projectiles are fin stabilized to reduce the effects of centrifugal force on the shaped charge jet formation. Though not common, there are spin-stabilized HEAT projectiles, but they require additional engineering consideration such as the "fluted" cone shown in Figure 4.35.

General identification features associated with HEAT projectiles include:

- **Appearance and materials:**
  - Heavy duty construction.
  - A break in the major diameter forward of the bourrelet.
  - Standoff spike (Figure 4.36).
  - A hollow ogive crimped or screwed to the body; may have spanner holes.
  - May have a tracer element on the base.

Figure 4.36 Russian HEAT projectiles with standoff spikes and piezoelectric crystal PIBD fuzing. Left: a 125mm tank fired projectile. Right: a 122mm artillery fired projectile. Note the claw like lugs protruding from the front face around the base of the standoff spikes. These are an effective anti-deflection measure. (Author's photograph.)

Figure 4.35 A 152mm U.S. HEAT projectile with a standoff ogive and a Point Initiating Base Detonating (PIBD) fuze with a piezoelectric crystal in the PI element. A unique feature of this munition is the fluted cone. (Author's photograph.)

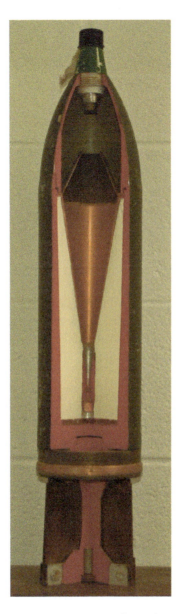

Figure 4.37 Russian 122mm HEAT projectile with a standoff ogive and a spit-back PIBD fuze. (Author's photograph.)

- **Markings:** A green, black, or gray body with yellow or black markings is common. Other colors, stamped or stenciled markings, and symbols may also be present.
- **Common fuze configurations:** BD, point initiating base detonating (PIBD) electric and PIBD mechanical fuzing (Figures 4.35–4.37). To function properly, all shaped charges must be initiated from the base

of the warhead. To address this requirement only BD, PIBD fuzing systems are used with HEAT projectiles. Due to the additional hazards associated with piezoelectric fuzing, assume all HEAT projectiles contain a PIBD (piezoelectric) fuze and that the BD element is present until proven otherwise.

- **General safety precautions** for HEAT projectiles include:
  - HE, frag, movement, jet.
  - Safety precautions for the fuze if present.

Note: Always assume that a HEAT projectile has a piezoelectric (PE) fuze until proven otherwise and include PE, electromagnetic radiation (EMR), and static.

3. Projectile, Guided: These are a few guidance system technologies used with projectiles. An example of an older technology is the US Copperhead (Figure 4.38), which incorporates laser guidance, where the new Excalibur uses satellite-guidance. Often confused for missiles, guided projectiles do not have a motor and are constructed with a heavy body design to withstand the forces associated with firing a projectile at extreme ranges. For example, in Afghanistan, an Excalibur projectile hit a target at a range of 36 km. Guided projectile designs most often fall into the HE or HEAT groups.

General identification features associated with guided projectiles include:

- **Appearance and materials:**
  - Heavy-duty construction.
  - High-quality machining of body, fins, screw heads, and section connections.
  - The presence of steerable fins consistent with missiles, but lacking motors or a means of attaching a motor.

- **Markings:** A green, black, or gray body with yellow or black markings is common. Other colors, stamped or stenciled markings, and symbols may also be present.

- **Common fuze configurations:** PD, BD, PIBD, or VT fuzing, which may not be visible.

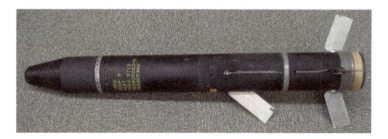

Figure 4.38 155mm, M712 U.S. Copperhead guided projectile. (Author's photograph.)

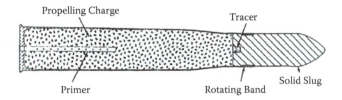

Propelling Charge

Tracer

Primer

Rotating Band

Solid Slug

Figure 4.39 Line drawing of a 90mm Armor Piercing-Tracer (AP-T). (From U.S. military TM.)

- **General safety precautions** for guided projectiles:
  - HE, frag, movement.
  - HEAT if applicable.
  - Safety precautions for the fuze if present.

Note: Fins may forcibly deploy, requiring an ejection safety precaution.

Note: Many guidance systems contain heavy metals and other toxic hazards. If a guidance system is damaged, adhere to the chemical safety precaution.

4. Projectile, Armor Piercing (AP): Was the first munition capable of defeating armored vehicles. Initial designs were nothing more than a hardened metal projectile that penetrated or spalled armor through kinetic impact. Later designs included explosive charges behind a hardened nose to enhance penetration and provide additional fragmentation against supporting infantry. Both designs have evolved over time and are still used today.

4a. Projectile, AP: Does not contain explosives and is designed to spall or penetrate armor with kinetic impact (Figures 4.39 and 4.40). New armor configurations rendered the original AP designs ineffective, which led to a new generation of AP projectiles. The armor-piercing discarding sabot (APDS) is a dart-shaped tungsten or depleted uranium (DU) penetrator of approximately 35mm to 40mm in diameter housed within a scooped-front sabot consisting of three separate parts (Figures 4.41 and 4.42). When an APDS is fired from a 105mm gun, the penetrator and sabot travel down the bore. Upon exiting, the three sabot parts separate from the subcaliber penetrator that is now propelled by the energy used to fire a 105mm projectile at a very

Figure 4.40 90mm AP-T. (Author's photograph.)

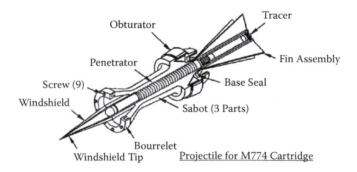

Figure 4.41 Line drawing of an Armor Piercing Discarding Sabot (APDS). (From U.S. military TM.)

high velocity. The result is a kinetic penetrator moving at speeds capable of effectively defeating new armor configurations.

General identification features associated with AP projectiles include:

- **Appearance and materials for AP projectiles:**
  - Heavy body construction that comes to a dull point.
  - May have a thin metal ballistic windshield over the nose to decrease resistance and increase velocity.
  - May or may not have hammer rings.
  - Multiple rotating bands or a very wide single band (Figure 4.40).
  - Tracer element on base.

Figure 4.42 Variety of Russian and American APDS projectiles. (Author's photograph.)

- **Appearance and materials for APDS projectiles:**
  - Solid, one-piece body that comes to a dull point.
  - Ribbed hammer rings on midbody (Figure 4.41 and the two projectiles on the right in Figure 4.42).
  - Fins for stabilization.
  - Fins larger in diameter than the penetrator.
  - Tracer element on base.

- **Markings:** A black or gray body with white or black markings is common. If a tracer element is present, a "T" designation may be present on the projectile. Other colors, stamped or stenciled markings, and symbols may also be present.
- **Common fuze configurations:** None.
- **General safety precautions** for AP projectiles include:
  - Movement.

4b. Projectile, Armor Piercing High Explosive (APHE): Like AP projectiles, an APHE is also designed to spall or penetrate armor with kinetic impact. However, an APHE also employs hammer rings on the body and an explosive charge in the base, both of which greatly intensify energy transmission (Figures 4.42 and 4.43).

General identification features associated with APHE projectiles include:

- **Appearance and materials:**
  - Solid body that comes to a dull point.
  - Hammer rings.
  - Multiple rotating bands or a very wide single band.
  - A crimped or pressed on ballistic windshield.
  - Tracer element on base.

- **Markings:** A green, black or gray body with yellow, white or black markings is common. If a tracer element is present, a "T" designation may be present on the projectile. Other colors, stamped or stenciled markings, and symbols may also be present.

Figure 4.43 Russian 152mm APHE-T. Note the 4 deep hammer-rings, which are a key identification feature. (Author's photograph.)

- **Common fuze configurations:** BD.
- **General safety precautions** for APHE projectiles include:
  - HE, frag, movement

5. Projectile, Anti-personnel (APERS) is designed to propel large amounts of shot, shrapnel, or flechettes against personnel in the open. There are three fundamental designs to address different ranges, two of which (1 and 2) have remained fundamentally unchanged for hundreds of years:

1. A projectile converts a field gun into an oversized shotgun as the outer casing strips away immediately upon exiting the bore dispersing the shot (Figure 4.44).

Figure 4.44 90mm U.S. M336 Canister projectile filled with barrel-shaped shot. (Author's photograph.)

Figure 4.45 The initial shrapnel design consisted of a cannonball filled with shrapnel, a bursting charge and time delay fuze. (Author's photograph.)

2. A classic "shrapnel" design most commonly employs a powder train time fuze (PTTF) that functions after a predetermined time. Upon fuze functioning, a flame travels down an internal tube to a black powder expelling charge in the base. The base is strongly constructed so that when the black powder initiates, the explosive force ejects the fuze adapter and propels the payload of shrapnel in a forward direction (Figures 1.2, 4.45, and 4.46).

3. In a more modern configuration, a specially designed Mechanical Time (MT) fuze set to meters versus time ensures a short-range airburst. The high-explosive burster disperses flechette, which are the dart-like modern version of shrapnel (Figures 4.47 and 4.48).

General identification features associated with APERS projectiles include:

- **Appearance and materials:**
  - Type 1:
    - A flat nose with no observable fuzing.
    - May have a tracer element.
    - Aluminum or sheet steel construction with longitudinal scores on the side of the body to allow easy opening.
    - **Markings:** May be painted black, green, or gray.
  - Type 2:
    - A PTTF fuze.
    - Steel construction with a solid body and a tracer element.
    - **Markings:** May be painted green, yellow, black, gray, or red.

Figure 4.46 A U.S. post-Civil War, Hotchkiss shrapnel projectile with internal spacers and a copper rotating band. (Courtesy of Dan Evers.)

- Type 3:
    - MT fuze with setting increments in distance.
    - Aluminum or sheet steel construction with longitudinal scores on the side of the body to allow easy opening.
- **Markings:** May be green or gray with yellow or black markings, which may include a row of white diamonds on the body.
- **Common fuze configurations:** MT and PTTF .
- **General safety precautions** for APERS projectiles include:
    - Movement.
    - HE, frag if applicable.
    - Safety precautions for the fuze if present.

Note: All time fuzes require a wait time (W/T) safety precaution.
Note: All MT fuzes require a cock striker (C/S) safety precaution.

6. Projectile, Dispenser and Improved Conventional Munition (ICM):
A dispenser is a hollow projectile body filled with a payload and sealed. At a predetermined time during flight a time fuze functions, initiating a

Figure 4.47 90mm U.S. M580 APERS-T filled with flechettes. Fitted with a special Mechanical Time (MT) fuze set to meters versus time to ensure a short-range air-burst, filling the air with well over 4,000 flechettes. (Author's photograph.)

low-explosive expelling charge that overcomes the pressed or threaded base or nose and ejects the payload from the projectile body. The term "improved conventional munitions" is reserved for dispensers containing HE or HEAT submunitions, which are covered in a later chapter. An ICM containing HE

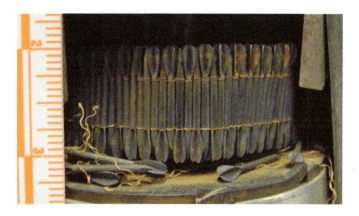

Figure 4.48 Close-up of flechette configuration within the projectile. (Author's photograph.)

submunitions, such as the M483 shown in Figures 4.49 and 4.50, is considered to be an HE projectile. After an ICM has functioned as designed and the submunitions have been ejected from the projectile, all that should remain is the empty projectile body. Once ejected, the payload from an ICM is categorized as a submunition as it has been separated from the projectile. Bursting smoke, burning smoke, and illumination dispenser-type projectiles will be covered later in this chapter. However, ICMs differ from these as they can dispense a payload, or with the addition of a booster, function upon impact as an enhanced fragmentation projectile.

Most dispensers eject from the base, but there are forward ejecting configurations. Forward ejecting models tend to have a flat nose that is crimped or pressed in place and a base fuze. Base-ejecting models have a solid base plate that is pressed or threaded in place and secured with shear pins (Figures 4.23 and 4.26).

General identification features associated with dispenser and ICM projectiles include:

- **Appearance and materials:**
  - Multi-piece body of steel, aluminum, and fiberglass.
  - Rotating band(s), gas-check bands, or an obturator ring.
  - A threaded or pressed-in base-ejecting projectile may have shear pins above the base, or where the tail boom attaches to the body. A crimped on or pressed-in nose may not have shear pins.
  - Spanner holes may be present in the base plate.
- **Markings:** The payload will define the color codes and markings. Stamped or stenciled markings and symbols may also be present.
- **Common fuze configurations:** MT, PTTF, and Electronic Time (ET) fuzing with an impact backup feature.

Figure 4.49 155mm U.S. M483 ICM, containing a payload of 88 M42 and M46 submunitions. (Author's photograph.)

- **General safety precautions** for dispenser and ICM projectiles include:
  - Movement, ejection.
  - Safety precautions for the fuze if present.

Note: Depending on the payload, HE, frag, jet, white phosphorus (WP), fire, and chemical may apply.

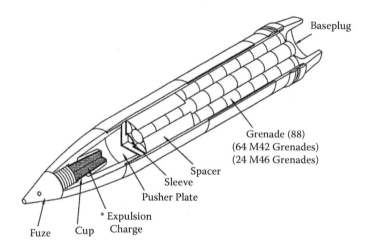

Baseplug

Grenade (88)
(64 M42 Grenades)
(24 M46 Grenades)

Spacer
Sleeve
Pusher Plate
* Expulsion
Fuze   Cup   Charge

Figure 4.50 Line drawing of a 155mm U.S. M483 ICM. (From U.S. military TM.)

Note: The payload is explosively deployed and offers a substantial ejection hazard.

7. Projectile, Smoke: Are designed to produce smoke for screening, marking targets, or destroying material with fire. There are two distinctly different types of smoke projectiles: bursting smoke and burning smoke. Additionally, projectiles that deploy riot control agents usually use burning smoke projectile designs and are also covered under this group. The key identification feature that will assist in differentiating between a burning smoke and riot control projectile are the color codes and markings.

7a. Projectile, bursting smoke, White Phosphorus (WP): There are two different types of bursting smoke projectiles that function in distinctly different manners:

Type 1: WP is sealed in the projectile by a burster adapter, which contains a high-explosive burster extending from the fuze well, down the center of the projectile. Upon impact with the ground, the PD fuze functions the burster, which detonates, breaking the projectile body into pieces while dispersing the jelly-like WP (Figures 4.51 and 4.52).

Type 2: Is a newer design in which a projectile dispenser is loaded with a hermetically sealed steel canister containing felt wedges impregnated with WP. At a predetermined time during flight, the MT or ET fuze functions, initiating the expelling charge that ejects the container and initiates a short delay. The delay functions the

Figure 4.51 60mm U.S. WP mortar with a similar internal design as Figure 4.52. **Note:** The ignition holes are between the fins, under the increment bags. (Author's photograph.)

burster running down the center of the container, breaking it apart and dispersing the WP felt wedges (Figures 4.53 and 4.54).

General identification features associated with bursting smoke projectiles include:

- **Appearance and materials:**
  - Type 1:
    - This is a solid, one-piece body of robust construction.
    - Rotating band(s), gas-check bands, or an obturator ring is present.
    - A burster adapter between the projectile body and fuze seals in the WP.
    - The adapter booster may have wrench flats or spanner holes.
    - On non-U.S. ordnance, top-down or side spanner holes on the adapter may indicate if the projectile is HE or WP (See for example Figure 20 in Chapter 7).
    - A solid base is common.
  - Type 2:
    - This is a multi-piece body of steel, aluminum, and fiberglass.
    - The rotating band is just forward of the break where the base ejects, which can be seen in Figures 4.53 and 4.54.

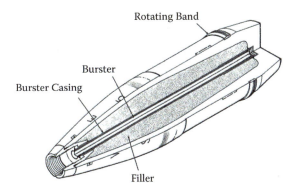

Figure 4.52 Line drawing of a 155mm U.S. WP projectile. (From U.S. military TM.)

Figure 4.54 Cutaway of Figure 4.27. Note that the projectile body and the internal container are both cutaway, exposing the inside frame where the felt wedges are stacked. (Author's photograph.)

Figure 4.53 155mm U.S. base-ejecting WP projectile. (Author's photograph.)

- **Markings:** A lime green or gray body with yellow and red markings is common. Other colors, stamped or stenciled markings, and symbols may also be present.
- **Common fuze configurations:**
  - Type 1: PD fuzing.
  - Type 2: MT or ET fuzing.

- **General safety precautions** for bursting smoke projectiles include:
  - HE, frag, movement, WP, fire.
  - Chemical if burning as WP smoke is toxic.
  - Safety precautions for the fuze if present.

Note: Add ejection for a base-ejecting model.

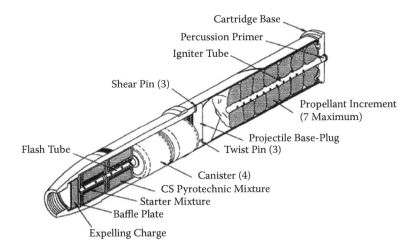

Figure 4.55 Line drawing of a 105mm, M629 CS projectile. (From U.S. military TM.)

7b. Projectile, Burning Smoke: Is a dispenser type projectile that ejects burning smoke canisters designed to produce colored smoke for signaling and screening. These projectiles do not contain high explosives, but do have a low-explosive expelling charge to eject the payload while also initiating the pyrotechnic smoke materials (Figure 4.55).

General identification features associated with burning smoke projectiles include:

- **Appearance and materials:** Include all of the construction features associated with dispenser type projectiles.
- **Markings:** A lime green or gray body with yellow and red markings is common. Other colors, stamped or stenciled markings, and symbols may also be present.
- **Common fuze configurations:** MT, PTTF, and ET fuzing.
- **General safety precautions** for burning smoke projectiles include:
  - Movement, fire, ejection.
  - Chemical if burning; HC, which produces white smoke, is toxic in field concentrations.
  - Safety precautions for the fuze if present.

Note: The payload is explosively deployed and offers a substantial ejection hazard.

7c. Projectile, Riot Control: Is a dispenser type projectile that ejects burning smoke canisters that produce CN, CS, CN1, and other pepper-like riot control substances (see Appendix B for full names of substances). These projectiles do not contain high explosives, but do have a low-explosive expelling charge to eject the payload while also initiating the riot control materials (Figure 4.55).

General identification features associated with riot control projectiles include:

- **Appearance and materials:** Include all of the construction features associated with dispenser type projectiles.
- **Markings:** A green or gray body with red and brown markings is common. Other colors, stamped or stenciled markings, and symbols may also be present.
- **Common fuze configurations:** MT, PTTF, and ET fuzing.
- **General safety precautions** for riot control projectiles include:
  - Movement, fire, ejection, and chemical.
  - Safety precautions for the fuze if present.

Note: The payload is explosively deployed and offers a substantial ejection hazard.

8. Projectile, Illumination: Is a dispenser type projectile that ejects a parachute-suspended, pyrotechnic candle used to illuminate an area at night. These projectiles do not contain high explosives, but do have a low-explosive expelling charge to eject the payload. The illumination candle usually ignites upon parachute deployment. Depending on the projectile size, candles can range in intensity from a few thousand to one million candlepower.

General identification features associated with illumination projectiles include:

- **Appearance and materials:**
  - All of the construction features associated with dispenser type projectiles.
  - Most configurations are nose fuzed, base ejecting (Figure 4.56).
  - Base-fuzed, front ejecting versions are less common. Note the MT fuze time increments on the base of Figure 4.57.

- **Markings:** A white, green, or gray body with black, red, or brown markings is common. The umbrella-like symbol seen in Figure 4.19 is commonly used to identify illumination munition from all ordnance categories. Other colors, stamped or stenciled markings, and symbols may also be present.
- **Common fuze configurations:** MT, PTTF, and ET fuzing. In some cases the ejection charge ignites the candle; if not, there will be a second fuze to accomplish this. Many configurations use parachute deployment to initiate a mechanical type fuze.

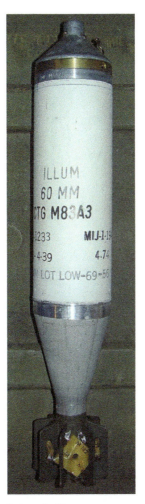

Figure 4.56 60mm, fin stabilized Illumination mortar with a Powder-Train-Time-Fuze (PTTF) on nose. (Author's photograph.)

- **General safety precautions** for illumination projectiles include:
  - Movement, fire, ejection.
  - Chemical if candle is burning.
  - Safety precautions for the fuze if present.

Note: The payload is explosively deployed and offers a substantial ejection hazard (Figure 4.58).

9a. Projectile, practice, with and without spotting charges: Are designed to be fired with the same ballistic characteristics as the live projectile it mimics, without the destructive effects. This includes subcaliber practice projectiles that are fired through modified weapon systems. Practice projectiles may be solid metal or contain inert filler such as powder dye, or a

Figure 4.57 60mm, fin stabilized Illumination mortar with base Mechanical Time (MT) fuze. Note time-set increments. (Author's photograph.)

spotting charge propelled by a substantial explosive charge initiated by a live fuze (Figure 4.59).

General identification features associated with practice projectiles include:

- **Appearance and materials:** Construction features associated with the projectile they are designed to imitate.
- **Markings:** Blue or black with white markings is common. The marking "TP" for target practice is found on many practice projectiles that contain no energetics or spotting charges. Other colors, stamped or stenciled markings, and symbols may also be present.

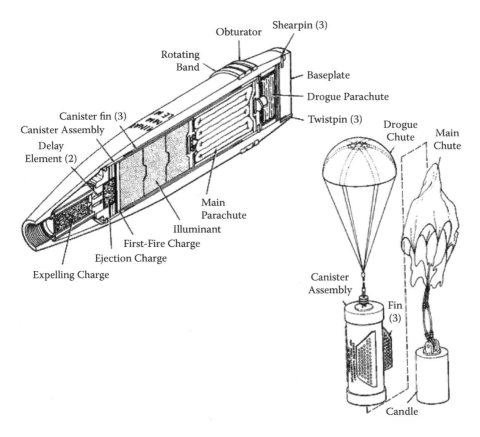

Figure 4.58 Line drawing of a 155mm illumination projectile, with deployment graphics. (From U.S. military TM.)

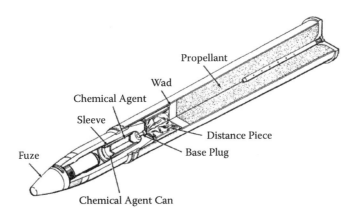

Figure 4.59 Line drawing of a 76mm U.S. MK167 "Practice" projectile containing a 4.1 ounce liquid FS spotting charge that is explosively dispersed by a live fuze. The component identified as "Chemical Agent" is FS. (From U.S. military TM.)

Figure 4.60 Naval drill round with wooden body. (Author's photograph.)

- **Common fuze configurations** depend on projectile type.
- **General safety precautions** for practice projectiles include:
  - Movement.
  - HE, frag, ejection when a spotting charge is present.
  - Observe all applicable safety precautions for the live projectile until positive identification is made.
  - Safety precautions for the fuze if present.

Note: Practice means "practice"—not "inert."

9b. Projectile, drill and dummy: Are not designed to be fired, but to be used for weapon system loading and unloading drills, or for display and contain no energetic materials whatsoever (Figure 4.60).

General identification features associated with drill and dummy projectiles include:

- **Appearance and materials:**
  - Construction features associated with projectile they are designed to imitate.
  - Older models made of wood.

- **Markings:** A gold or blue body with white markings is common. Other colors, stamped or stenciled markings, and symbols may also be present.
- **Common fuze configurations:** None.
- **General safety precautions** for drill and dummy projectiles include:
  - Movement, until positive identification is made.

Closing

All ordnance including practice projectiles are inherently dangerous. Until proven otherwise, always consider a projectile to be in a hazardous condition.

Ordnance Category— Grenades: Hand, Rifle, and Projected

5

The young man knows the rules, but the old man knows the exceptions.

Oliver Wendell Holmes

Introduction

Under the category "grenade" there are three distinctly different category types that are broken down in Appendix A, logic tree 2:

1. Hand grenade
2. Rifle grenade
3. Projected grenade

There have been thousands of different grenade designs manufactured over the years. This chapter focuses on establishing the category-type and group to which an unknown grenade belongs. Yet with so many designs, many of which are culturally influenced, it is impossible to cover every grenade. The rules associated with shapes, construction features, materials, and functional designs certainly apply. But oftentimes, detecting an exception to an established rule offers the key to positively identifying an unknown munition.

Hand grenades date back to the 1600s when softball-sized cast iron balls filled with black powder and a burning fuze were thrown by specially trained assault troops known as "grenadiers." By the beginning of WWI hand grenades did not require specialized training and could be thrown by most soldiers. Then a new type of grenade launched from a rifle and designed to address the range between a hand grenade and a mortar was fielded. Rifle grenades would be the grenadiers' primary weapon for many years. By the 1950s a more efficient delivery system was designed for grenadiers that did not rely on a rifle. Fired from what appears to be an oversized shotgun, projected grenades possess all of the characteristics of a projectile. But as they were designed to be deployed by grenadiers, they are classified under the grenade category, which is an important caveat when researching an unknown munition.

There are numerous hand, rifle, and projected grenade designs that all consist of a fuze, body, or warhead filled with a wide variety of energetic or inert materials. Depending on the design, most practice grenades contain inert filler, but some incorporate a variety of dangerous fillers, requiring all to be treated as extremely hazardous until proven otherwise.

Hand Grenades

Hand grenades are manually armed and thrown. With a variety of fuzes available, hand grenades may function after a predetermined time or impact with a target. Some are equipped with extended handles for leverage or to enclose a means of orientation such as a parachute that deploys after being thrown. Due to the number of grenades manufactured, their availability to almost every soldier, and relatively small size, grenades are commonly recovered outside military control.

Key Identification Features for Hand Grenades

Ranging in size from a golfball to a 1 L bottle, the manner in which they are deployed offers the best identification features. For example, hand grenades designed in countries with popular sports involving the throwing of a ball—such as baseball and cricket—tend to produce somewhat round or oval-shaped grenades that can be thrown like a ball (Figure 5.1). Soldiers

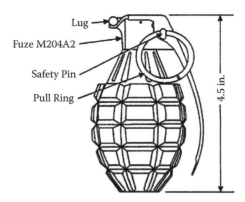

Figure 5.1 One of the most recognizable designs, the U.S. MK2 fragmentation hand grenade, commonly referred to as a "pineapple" grenade. (From U.S. military training manual [TM].)

Figure 5.2 One of the two most recognized designs, the German models 24 and 29 stick grenades, commonly referred to as the "potato masher" grenade. (Author's photograph, courtesy of Greg Everett.)

from countries without such popular sports may not have developed good throwing fundamentals in their youth. In order to increase the range these soldiers can throw a hand grenade, designs with features such as a stick are used to increase throwing leverage (Figure 5.2).

Hand grenades can be quite small and are usually not larger than what the average person will be able throw a distance greater than the hazards the grenade possesses.

The Seven-Step Practical Process Applied to Hand Grenades

Examples of different designs, features, color codes, markings, and construction features are provided throughout this chapter.

Step 1: Approach and initial interrogation. Attempt to identify a munition at a distance with the use of binoculars. If an approach is made, measurements must be taken of the major diameter and overall length of the body, fuze, and all unique features.

All findings, including measurements, stamped data, color codes, markings, key identifying features, and any possible damage, are documented and the grenade is photographed. In addition to the overall configuration, there are three features that will greatly assist in answering steps 2, 3, 5, and 7:

1. The diameter.
2. The overall length.
3. The design and materials used on the fuze.

Step 2: Determine fuze type and condition. Many hand grenade fuzes are armed when a safety pin, clip, cap, or string is removed. If a hand grenade has been deployed or any safety device has been removed, the fuze is considered to be armed (step 5). If a fuze is damaged, pins have been removed, or any alterations have been made to the munition, it is considered armed. If visible, measurements for the fuze are taken separately from the munition.

Step 3: Determine ordnance category. This category covers grenades designed to be hand thrown.

Step 4: Determine ordnance group. Identifying characteristics associated with each hand grenade group will be covered throughout this chapter, many of which are consistent with the grouping characteristics of other categories.

Note: In reference literature the term "body" is commonly used to describe the primary component of hand grenades. Depending on the reference, this section of a grenade can be interchangeably described as the warhead or the body.

Step 5: Determine if the munition was deployed. If the grenade is damaged by impact or if a pin, clip, cap or string was removed, assume the munition was deployed (step 2).

Step 6: Determine safety precautions that apply to the munition. The safety precautions for the hand grenade groups are covered in this chapter. Chapter 3 addresses the safety precautions associated with various fuzes.

Note: Adhere to all safety precautions that apply.

Step 7: Identify the munition. Apply the totality of all construction characteristics and other identifying features to determine the group to which a hand grenade belongs and, if feasible, positively identify the munition and all possible fuzing configurations.

Note: Some submunitions are referred to as "grenades" in research literature. Chapter 9 will provide an example of this.

Groups

The category "hand grenade" includes these groups:

- a. Fragmentation (frag).
- b. Blast.
- c. HEAT.
- d. Bursting smoke.
- e. Burning smoke.
- f. Illumination.
- g. Incendiary.
- h. Riot control.
- i. Practice.

1a. Hand grenade, Fragmentation ("frag"): Usually contains between 1 and 8 oz (28 and 227 g) of high explosives in a serrated or microengraved metal body designed to fragment. Some are small (Figure 5.3) and many have a "pineapple" appearance (Figures 5.4 and 5.5). Fragmentation grenades are classified as defensive grenades and may contain incendiary materials to provide an anti-material effect. Some designs include a stick while others have a thin plastic outer body covering the fragmentation liner, which many be made of a plastic-like material embedded with single or multiple layers of metal fragmentation. Many of these grenades resemble those used by paintball enthusiasts or children's toys (Figures 5.6–5.8). Grenade bodies come in many shapes, sizes, and configurations with many having internal serrations and a smooth outer appearance (Figure 5.9).

General identification features associated with fragmentation hand grenades include:

- **Appearance and materials:**
  - Externally or internally serrated iron or steel body of heavy construction.
  - An external spring-like fragmentation sleeve (Figure 5.10).
  - A plastic or sheet-metal outer cover concealing a fragmentation liner.
  - An external fuze with a "spoon" held in place by a safety pin.
  - "Stick grenades" with wooden or light metal handles incorporate an internal fuze with a cap covering a ring attached to a string.

- **Markings:** A green body with yellow or red markings and a black body with red markings are common. Other colors, stamped or stenciled markings, and symbols may also be present on the grenade.

Figure 5.3 German "Egg-Type" hand grenades. (Courtesy of Didzis Jurcins.)

Figure 5.4 British "Mills Bomb." There are many versions of this grenade; this is a No 36 MK1 variant with a C/S reverse-acting fuze. (Courtesy of Dan Evers.)

- **Common fuze configurations:** Pyrotechnic delay fuze, that is direct-armed prior to being thrown is most common. Direct-armed electrical, and mechanical impact-initiated fuzes are also employed.
- **General safety precautions** for fragmentation hand grenades include:
  - HE, frag, movement.
- Safety precautions for the fuze if present.

Figure 5.5 Russian F-1 grenade with a C/S reverse-acting fuze. (Courtesy of Dan Evers.)

Figure 5.6 Stick grenade with a pull-friction fuze. (Courtesy of Dan Evers.)

Figure 5.7 Austrian HDGR 73. The plastic shell covers an internal fragmentation liner. (Courtesy of Dan Evers.)

Figure 5.8 Japanese Type 97. After the pin is removed, the fuze is struck to initiate the pyrotechnic delay prior to being thrown. (Author's photograph.)

Figure 5.9 Austrian HGR 85. The smooth outer body may indicate internal serrations designed to maximize fragmentation. (Courtesy of Dan Evers.)

Figure 5.10 Belgian NR 7/8 with removable external fragmentation sleeve, which allows it to be deployed as a fragmentation or blast grenade. (Author's photograph.)

1b. Hand grenade, Blast: Usually contains between 1 and 8 oz (28 and 227 g) of HE housed in a soft body designed to produce blast and overpressure effects without fragmentation. The basic configuration of a blast grenade is the same as that of a frag grenade without the fragmentation liner (Figure 5.10). General identification features associated with blast hand grenades include:

- **Appearance and materials:**
  - Lightweight materials such as thin aluminum, cardboard, tarpaper, or plastic (Figures 5.10–5.14).

Figure 5.11 French OF-37 with a very thin metal body. (Courtesy of Didzis Jurcins.)

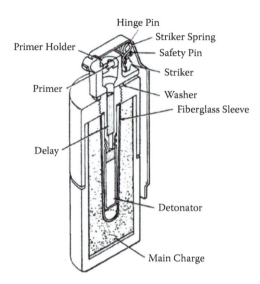

Figure 5.12 U.S. MK 3. The body is made of a tarpaper-like material. (From U.S. military TM.)

- A configuration able to accommodate a fragmentation sleeve (Figure 5.10, with liner removed).
- An external fuze with a "spoon" held in place by a safety pin.

- **Markings:** A black body with yellow markings or a green body with yellow markings is common. Other colors, stamped or stenciled markings, and symbols may also be present on the grenade.

Figure 5.13 Cutaway of a Russian RG4 grenade with an internal all-way-acting fuze. (Author's photograph.)

Figure 5.14 Marking, in English, on a Russian RG4 grenade. (Author's photograph.)

- **Common fuze configurations:** A pyrotechnic delay fuze that is direct-armed prior to being thrown is most common. Direct armed electrical, and mechanical impact-initiated fuzes are also employed.
- **General safety precautions** for blast hand grenades include:
  - HE, frag, movement.
  - Safety precautions for the fuze if present.

Note: The frag safety precaution is taken in order to address secondary fragmentation effects.

Note: Some fuzes used with blast grenades are designed to "pop" off prior to the grenade functioning to reduce fragmentation.

1c. Hand grenade, High-Explosive Anti-Tank (HEAT): Employs a shaped charge to defeat armor and other hardened targets with explosive charges ranging from 6 to 24 oz (171 to 680 g). In order to function properly, the grenade must have a means of stabilization to ensure proper orientation, a conically shaped charge, and a standoff to allow the shaped charge to function properly (Figures 5.15 and 5.16). All of these construction requirements provide unique features that assist in identifying a HEAT grenade. These grenades are not common and the United States has not fielded a viable HEAT hand grenade, but many countries have.

General identification features associated with HEAT hand grenades include:

- **Appearance and materials:**
  - A break in the major diameter on the body where the explosive charge, cone, and standoff meet (Figure 5.15).

Figure 5.15 Russian RKG-3, HEAT grenade. Note the break in the middle of the body. The section opposite the handle is hollow to provide standoff for the shaped charge. (Courtesy of Tom Conte.)

Figure 5.16 Cutaway of an RKG-3 handle exposing the fuze. (Courtesy of Tom Conte.)

- A handle, fins, parachute, cloth covered spring, or other means of orientation (Figure 5.16).
- Light sheet-metal or plastic construction.
- May have a plastic body that opens like a clamshell to provide orientation during deployment.
- A smooth outer appearance.
- Partially or completely internal fuze.
- **Markings:** A green or black body with yellow, red, or black markings is common. Other colors, stamped or stenciled markings, and symbols may also be present on the grenade.
- **Common fuze configurations:** Point-Initiated Base-Detonating (PIBD) and Base-Detonating (BD) fuzing that is direct-armed prior to being thrown is most common. If present, the Point Initiating (PI) element may be observable in the form of a plunger-like protrusion.

- **General safety precautions** for HEAT hand grenades include:
  - HE, frag, movement, jet.
- Safety precautions for the fuze if present.

1d. Hand grenade, Bursting Smoke: Is designed to mark targets or destroy material with dense white smoke and fire. Bursting smokes contain White Phosphorus (WP) and an HE bursting charge that fractures the body explosively spreading the WP filler.

General identification features associated with bursting smoke hand grenades include (Figure 5.17):

- **Appearance and materials:**
  - Size of a soda can or larger.
  - Heavy body construction that may have external serrations.

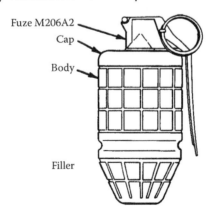

Fuze M206A2

Cap

Body

Filler

Figure 5.17 U.S. M34 bursting smoke. The groove just above the tapered base is to affix the grenade to a rifle grenade tail boom. (From U.S. military TM.)

- An external fuze with a "spoon" held in place by a safety pin.
- Some designs resemble burning smokes in shape and size, but lack the telltale emission holes or crimped edges (Figure 5.18).

- **Markings:** A lime green body with yellow and red markings is common. Other colors, stamped or stenciled markings, and symbols may also be present.
- **Common fuze configurations:** A pyrotechnic delay fuze that is direct-armed prior to being thrown is most common. Direct armed electrical, and mechanical impact-initiated fuzes are also employed.
- **General safety precautions** for bursting smoke hand grenades include:
  - HE, frag, movement, WP, fire.
  - Chemical for smoke if WP is burning.
  - Safety precautions for the fuze if present.

1e. Hand grenade, Burning Smoke: Contains pyrotechnic mixtures that burn to produce smoke in a variety of colors for signaling or screening. A small flame and hot gas are produced with the colored smoke, but the grenade body remains intact after functioning.

General identification features associated with burning smoke hand grenades include (Figures 5.18 and 5.19):

- **Appearance and materials:**
  - Size of a soda can or larger.
  - Light sheet metal construction with smooth outer appearance.

Figure 5.18 Line drawing of a U.S. M18 burning smoke grenade with four emission holes on top. (From U.S. military TM.)

Figure 5.19 U.S. M83 smoke grenade with one emission hole on bottom. The tape will tear through, with the grenade functions allowing the smoke to exit. (Author's photograph.)

- An external fuze with a "spoon" held in place by a safety pin.
- Telltale emission holes for smoke to escape on top or bottom (Figures 5.18 and 5.19).
- Rolled or crimped edges.

- **Markings:** A green body with white markings is common. The color of the top and a band on the body may indicate the color of smoke produced. Other colors, stamped or stenciled markings, and symbols may also be present on the grenade.
- **Common fuze configurations:** A pyrotechnic delay fuze that is direct-armed prior to being thrown is most common.
- **General safety precautions** for burning smoke hand grenades include:
  - Movement, fire.
  - Chemical: HC (hexachlorethane) white smoke is toxic in field concentrations.
  - Safety precautions for the fuze if present.

1f. Hand grenade, Illumination ("Illum"): Contains pyrotechnic mixtures that burn intensely to illuminate an area or for signaling. The example provided in Figure 5.20 consists of two sheet-metal cups pressed together

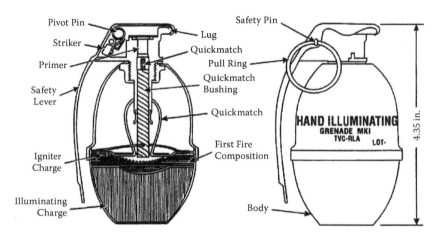

Figure 5.20 U.S. MK1 grenade, which burns at approximately 55,000 candle-power, is capable of illuminating an area 200 m in diameter and poses a serious fire hazard. (From U.S. military TM.)

and sealed. When the fuze functions, the two halves are separated and the pyrotechnic mixture is ignited.

General identification features associated with illumination hand grenades include:

- **Appearance and materials:**
  - Light sheet-metal construction with smooth outer appearance.
  - An external fuze with a "spoon" held in place by a safety pin.
  - A seam at approximately midbody.

- **Markings:** A white or green body with black or white markings and a white band is common. Other colors, stamped or stenciled markings, and symbols may also be present on the grenade.
- **Common fuze configurations:** A pyrotechnic delay fuze that is direct-armed prior to being thrown is most common. Many designs do not incorporate a delay and function immediately after being thrown.
- **General safety precautions** for illumination hand grenades include:
  - Movement, fire.
  - Chemical if candle is burning.
  - Not looking directly at a burning candle.
  - Safety precautions for the fuze if present.

1g. Hand grenade, Incendiary: Usually contains thermite or thermate mixtures that burn at approximately 4000°F to destroy equipment by melting or burning.

Figure 5.21 U.S. AN-M14 incendiary hand grenade, which contains 1.6 lb of TH3 thermate. (Author's photograph.)

General identification features associated with incendiary hand grenades include (Figure 5.21):

- **Appearance and materials:**
  - Size of a soda can or larger.
  - Light sheet-metal construction with smooth outer appearance.
  - May have rolled or crimped edges.
  - Similar size and configuration as many burning smoke grenades, but lacking the telltale emission holes.
  - An external fuze with a "spoon" held in place by a safety pin.

- **Markings:** A red body with black markings is common. Other colors, stamped or stenciled markings, and symbols may also be present on the grenade.
- **Common fuze configurations:** A pyrotechnic delay fuse that is direct-armed prior to being thrown is most common. Many designs do not incorporate a delay and function immediately after being thrown.
- **General safety precautions** for incendiary hand grenades include:
  - Movement, fire.
  - Chemical if grenade is burning.
  - Safety precautions for the fuze if present.

1h. Hand grenade, Riot Control: CN (chloroacetophenone) and CS (chlorobenzalmalononitrile) are the most common riot control agents used with military ordnance. There are multiple variations of these chemical compositions, such as CN1, but all are pepper-like substances that provide similar

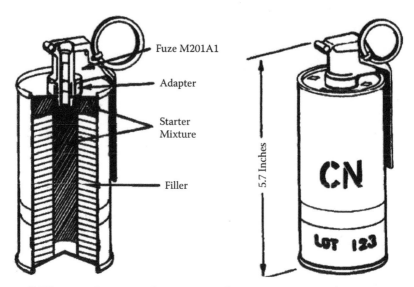

Figure 5.22 Line drawing of a U.S. M7 burning, riot-control hand grenade. (From U.S. military TM.)

effects. There are two common designs of riot control hand grenades, which will be covered together to decrease the possibility of confusing the two:

- Burning: Deployed in the same body configuration as many burning smoke grenades (Figure 5.22).
- Bursting: It is important to note that this design does not produce smoke through burning. Conversely, it contains an explosive burster that fragments the body, dispersing a powdered agent (Figures 5.23 and 5.24).

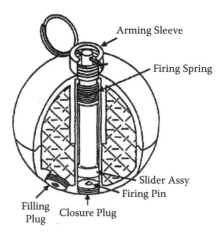

Figure 5.23 Line drawing of a U.S. M25A1 bursting, riot-control hand grenade. (From U.S. military TM.)

Figure 5.24 M25A1 bursting, riot-control hand grenade. (Author's photograph.)

General identification features associated with riot control hand grenades include:

- **Appearance and materials:**
  - Burning: These have the exact same body as many burning smoke grenades, including the crimped ends, and telltale emission holes. An external fuze with a "spoon" held in place by a safety pin is common (Figures 5.18 and 5.22).
  - Bursting: These are usually round, about the size of a baseball or softball. The body may be hard rubber, bakelite plastic, or other hardened material. A filler plug and low-profile fuze may be employed.

- **Markings:**
  - Burning: A gray body with red bands and markings is common. Color codes or markings are required to differentiate this type of riot control grenade from colored smoke grenades.
  - Bursting: Body colors vary depending on the materials used. For example, rubber bodies may be black and a bakelite plastic body may be brown (Figure 5.24). But a gray body with red bands or markings is common.

Note: For both riot-control grenades, there may be other colors, stamped or stenciled markings, and symbols present on the body.

- **Common fuze configurations:** A pyrotechnic delay fuze that is direct-armed prior to being thrown is most common. Many designs do not incorporate a delay and function immediately after being thrown.

- **General safety precautions** for riot control hand grenades include:
  - Burning: Movement, fire, and chemical.
  - Bursting: HE, frag, movement, and chemical.
  - Safety precautions for the fuze if present.

Note: Bursting type riot-control hand grenades contain a robust burster charge that fragments the body.

1i. Hand grenade, practice: Some practice hand grenades are designed to resemble a live grenade while others are designed to look and function in the same manner as a live grenade, but lack the full energetic filler of the live munition. "Practice" does not mean safe or inert, and many practice grenades contain significant hazards, including live fuzing and pyrotechnic spotting charges.

General identification features associated with practice hand grenades include (Figures 5.25 and 5.26):

- **Appearance and materials:**
  - Same shape, weight, and appearance as the live grenade it is designed to resemble.
  - An external fuze with a "spoon" held in place by a safety pin.

- **Markings:** A blue body with brown markings and a brown band is common. However, black bodies with white markings, yellow

Figure 5.25 South Korean K417 practice grenade with yellow spotting charge (note damage to fuze housing). Practice grenades are easily mistaken for paint-ball grenades or children's toys. (Author's photograph.)

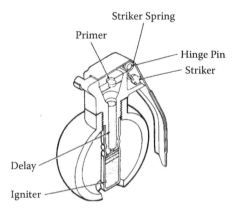

Figure 5.26 U.S. M69 practice grenade with a live fuze. The hole in the base of the empty body allows the energy from the fuze to vent. (From U.S. military TM.)

bodies with black markings, and other color combinations are also common.

- **Common fuze configurations:** A pyrotechnic delay fuze that is direct-armed prior to being thrown is most common.
- **General safety precautions** for practice hand grenades vary greatly depending on the model and fuze, but can include:
  - Movement
  - All applicable safety precautions for the live grenade until positive identification is made.
  - Safety precautions for the fuze if present.

Rifle Grenades

Rifle grenades were devised to cover the tactical range beyond hand grenades. Designs vary from a hand grenade fixed to a simple stabilizer tube assembly to complex designs. With a variety of configurations and fuzes available, rifle grenade fuzes may be manually armed by the removal of a pin, involve a multi-faceted arming sequence that transpires during flight, or a combination of both. There are three primary designs used to deploy rifle grenades munitions:

1. The gas produced by the firing of a blank cartridge propels the rifle grenade.
2. The gas produced by the firing of a bullet, which "passes through" the grenade capturing the gas behind it, propels the rifle grenade.
3. The gas produced by the firing of a bullet, as well as the bullet itself, propels the rifle grenade when the bullet is caught in a "bullet trap" in the forward end of the stabilizer tube assembly.

Figure 5.27 Stabilizer tube assemblies. Note the lack of ignition holes on or above the fins that are not the same diameter as the widest section of the body. Compare tail-booms with Figures 4.17 and 4.18. (Author's photograph.)

Key Identification Features

Rifle grenades usually consist of a body, fuze, energetic material, chemical or inert filler, and a stabilizer tube assembly. But it is the manner in which they are deployed that offers the best identification features. Most rifle grenades include a hollow stabilizer tube assembly that lacks the emission holes of a mortar or the venturi of a rocket (Figures 5.27–5.29). The most effective means of identifying a munition as a rifle grenade may be not what is present, but rather what is not.

Figure 5.28 Stabilizer tube openings. Note the shape of the opening and lack of a venturi or ignition holes on or above the fins. (Author's photograph.)

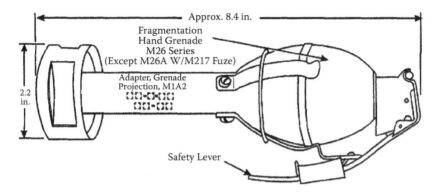

Figure 5.29 Line drawing of a MK26 Fragmentation Grenade mounted on an M1-Series "Grenade Projection Adapter," which converts a hand grenade into a rifle grenade. (From U.S. military TM.)

Rifle Grenade Sections

Stabilizer tube assembly. Sometimes referred to as a tail-boom, this section fits over the barrel of the rifle. It contains fins toward the base and attaches to the warhead or body at the forward end. Depending on the design, a bullet trap may be in the forward end and a fuze may be between the stabilizer tube assembly and the warhead or body. Not all rifle grenades have a stabilizer tube (Figures 5.30 and 5.31).

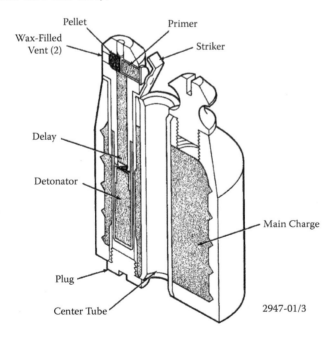

Figure 5.30 Line drawing of a U.S. MK-1, which is a copy of a French WWI era Vivien Bessiere (VB) Rifle Grenade. (From U.S. military TM.)

Figure 5.31 The Vivien Bessiere (VB) Rifle Grenade design was used by many countries throughout WWI. (Author's photograph).

Warhead or body. Oftentimes these terms are used interchangeably. For rifle grenades, the body is the primary component of the munition, and the warhead is the section that contains chemical agents, high explosives, or other energetic materials.

The Seven-Step Practical Process Applied to Rifle Grenades

Examples of different designs, features, color codes, markings, and construction features are provided throughout this section.

Step 1: Approach and initial interrogation. At a minimum, measurements must be taken of the major diameter and overall length of the rifle grenade. All findings to include measurements, color codes, markings, key identifying features, and damage are documented and the grenade is photographed. In addition to the overall configuration, there are three features that will greatly assist in answering steps 2, 3, 5, and 7:

1. The diameter.
2. The overall length.
3. A hollow stabilizer tube assembly lacking a venturi and the ignition holes commonly associated with mortars and recoilless rifle projectiles.

Step 2: Determine fuze type and condition. Many rifle grenade fuzes are armed when a safety pin is removed; others must be fired to arm. If a rifle grenade has been deployed, the fuze is considered to be armed (step 5). If a fuze is damaged or any alterations have been made to the munition, it is considered armed. If visible, measurements for the fuze are taken separately from the munition.

Step 3: Determine ordnance category. This category covers grenades designed to be fired from a rifle.

Step 4: Determine ordnance group. Color codes and markings vary from country to country and some fundamentals were addressed in Chapter 2. Under each group in this chapter, the common design and color configurations will be provided.

Step 5: Determine if the munition was deployed. If the rifle grenade is damaged by impact or if a pin was removed, assume the munition was deployed.

Step 6: Determine safety precautions that apply to the munition. The safety precautions for the rifle grenade groups are covered in this chapter. Chapter 3 addresses the safety precautions associated with various fuzes.

Note: Adhere to all safety precautions that apply.

Step 7: Identify the munition. Identify the rifle grenade as well as all possible fuzing configurations.

Groups

The category rifle grenade includes these groups:

    a. Fragmentation.
    b. HEAT.
    c. Bursting smoke.
    d. Burning smoke (includes colored smoke and riot control).
    e. Illumination.
    f. Practice.

2a. Rifle grenade, Fragmentation ("Frag"): Usually contain between 1 and 8 oz (28 and 227 g) of HE housed in a serrated or microengraved metal body designed to fragment. Incendiary materials may be added to provide an anti-material effect.

General identification features associated with fragmentation rifle grenades include (Figures 5.29–5.31):

- **Appearance and materials:**
    - ID features consistent with a fragmentation hand grenade with the addition of a stabilizer tube assembly.
    - Depending on design, external or internal fuze.
    - Hollow stabilizer tube assembly lacking a venturi or the ignition holes.
    - Grenade bodies with internal serrations but possibly a smooth outer appearance.

- Some designs with a thin plastic body covering a fragmentation liner and resembling those used by paintball enthusiasts or children's toys.

- **Markings:** A green or black body with yellow or red markings is common. Other colors, stamped or stenciled markings, and symbols may also be present on the body, fuze, or stabilizer tube assembly.
- **Common fuze configurations:** Pyrotechnic delay, PD, and BD fuzes that are direct-armed or partially armed prior to deployment or initiated during deployment are common. Direct-armed electrical, and mechanical impact-initiated fuzes are also employed.
- **General safety precautions** for fragmentation rifle grenades include:
  - HE, frag, movement.
  - Safety precautions for the fuze if present.

2b. Rifle grenade HEAT: Employs a shaped charge to defeat armor and other hardened targets. In order to function properly, the rifle grenade has a stabilizer tube assembly to ensure proper orientation, a conically shaped charge, a standoff to allow the shaped charge to function properly, and an explosive charge. All of these construction requirements provide unique features that assist in identifying a HEAT rifle grenade. HEAT rifle grenades offer an infantryman an anti-armor capability and reasonable standoff distance. Unlike HEAT hand grenades, which are uncommon, HEAT rifle grenades are employed by many countries.

General identification features associated with HEAT rifle grenades include:

- **Appearance and materials:**
  - A break in the major diameter of the warhead section where the explosive charge, cone, and standoff meet (Figures 5.32 and 5.33).
  - Depending on the design, partially external or completely internal fuze.
  - Hollow stabilizer tube assembly lacking a venturi or ignition holes.
- **Markings:** A green or black body with yellow, black, and red markings is common. Other colors, stamped or stenciled markings, and symbols may also be present on the body, fuze, or stabilizer tube assembly.
- **Common fuze configurations:** BD and PIBD (spitback), PIBD piezoelectric (PE) fuzes that are direct-armed or partially armed prior to firing and complete the arming process during flight are common.
- **General safety precautions** for HEAT rifle grenades include:
  - HE, frag, movement, jet.
  - Safety precautions for the fuze if present.

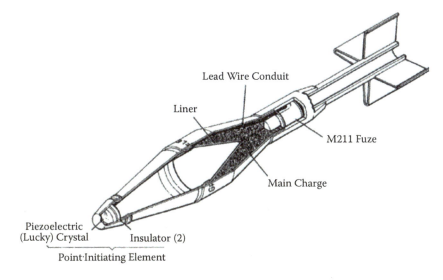

Figure 5.32 Line drawing of a U.S. M31, HEAT, rifle grenade, which contains a PE crystal PIBD fuze. (From U.S. military TM.)

Figure 5.33 German, WWII era, spin-stabilized, Grose Panzer Gewehr grenade (HEAT rifle grenade). (Courtesy of Didzis Jurcins.)

Note: Always assume that a HEAT rifle grenade has a PE fuze until proven otherwise and include PE, electromagnetic radiation (EMR), and static.

2c. Rifle grenade, bursting smoke: Designed to mark targets with dense white smoke or destroy material with fire. Bursting smoke rifle grenades contain WP and an HE bursting charge that fractures the body while explosively spreading the WP filler.

General identification features associated with bursting smoke rifle grenades include (Figure 5.34):

- **Appearance and materials:**
  - Light sheet-metal construction with smooth outer appearance.
  - Depending on design, partially external or completely internal fuze.
  - Some designs resembling burning smokes in shape and size, but lacking the telltale emission holes.
  - Hollow stabilizer tube assembly lacking a venturi or ignition holes.
  - A stabilizer assembly attached to a hand grenade such as the U.S. M34 (Figure 5.17).
- **Markings:** A lime green body with yellow markings is common. Other colors, stamped or stenciled markings, and symbols may also be present.
- **Common fuze configurations:** BD direct armed or partially armed fuzes that complete the arming process during flight are common.
- **General safety precautions** for bursting smoke rifle grenades include:
  - HE, frag, movement, WP, fire.
  - Chemical for smoke if WP is burning.
  - Safety precautions for the fuze if present.

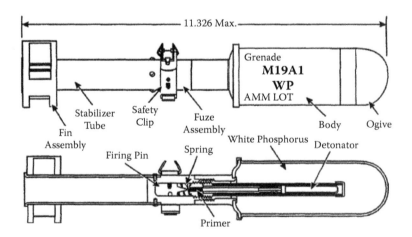

Figure 5.34 Line drawing of a U.S. M19A1 bursting smoke rifle grenade. (From U.S. military TM.)

2d. Rifle grenade, burning smoke, colored smoke and **riot control:** Contain pyrotechnic mixtures that burn to produce smoke in a variety of colors for signaling or screening. The rifle grenade body remains intact after functioning. This configuration is also used to deploy riot control agents, but as these rifle grenades are not very common, they are mentioned here versus being afforded their own section.

General identification features associated with burning smoke rifle grenades include (Figure 5.35):

- **Appearance and materials:**
  - Light sheet-metal construction with smooth outer appearance.
  - Most commonly, a completely internal fuze.
  - Telltale emission holes on bottom for smoke to escape.
  - Rolled or crimped edges.
  - Hollow stabilizer tube assembly lacking a venturi or ignition.

- **Markings:** A gray or green body with white or yellow markings is common. The color of the bottom of the body where it meets the stabilizer tube assembly or a band on the body may indicate the color of smoke produced. Riot-control rifle grenades may have a gray body with red bands or markings. Other colors, stamped or stenciled markings, and symbols may also be present.

- **Common fuze configurations:** BD, direct armed, or partially armed fuzes that complete the arming process during flight are common.

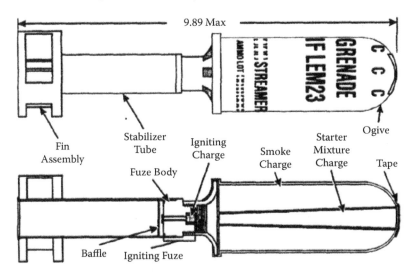

Figure 5.35 Line drawing of a U.S. M23 series burning smoke rifle grenade. (From U.S. military TM.)

- **General safety precautions** for burning smoke rifle grenades include:
  - Movement, fire.
  - Chemical if grenade is burning.
  - Safety precautions for the fuze if present.

2e. Rifle grenade, Illumination ("Illum"): Contains pyrotechnic mixtures that burn intensely to illuminate an area or for signaling. Upon firing, a pyrotechnic delay fuze is ignited and functions after a preset time. The fuze functions, and an expelling charge ejects the candle from the rifle grenade body. Depending on design, a parachute may or may not be present. If a parachute is present, it will deploy and the candle will ignite and burn as it floats toward the ground (Figure 5.36). If a parachute is not present, the candle mixture will burn while falling quickly toward the ground.

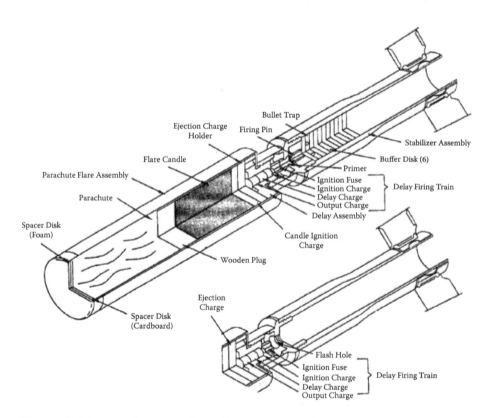

Figure 5.36 Line drawing of an illumination rifle grenade that is propelled and ignited by the firing of a bullet into the bullet trap in its base. (From U.S. Military TM.)

General identification features associated with illumination rifle grenades include (Figure 5.36):

- **Appearance and materials:**
  - Light sheet-metal construction with smooth outer appearance.
  - Internal fuzing that may have external time increments if the fuze is observable.
  - Rolled or crimped nose cap or base.
  - Hollow stabilizer tube assembly lacking a venturi or ignition holes.

- **Markings:** White or green body with black or white markings and a white band are common. The umbrella-like symbol seen in Figure 4.19 is commonly used to identify illumination munitions from all categories.
- **Common fuze configurations:** Mechanical Time (MT) and PTTF. In many cases the ejection charge ignites the candle; if not, there will be a second fuze to accomplish this. Many configurations use the parachute deployment to initiate a mechanical type fuze.
- **General safety precautions** for illumination rifle grenades include:
  - Movement, ejection, fire.
  - Chemical if candle is burning.
  - Safety precautions for the fuze or fuzes if present.

1f. Rifle grenade, practice: Some practice rifle grenades are designed to resemble a live grenade while others are designed to look and function in the same manner as a live grenade, but lack the full energetic filler of the live munition. "Practice" does not mean safe or inert, and many practice rifle grenades contain significant hazards, including live fuzing and pyrotechnic spotting charges.

General identification features associated with practice rifle grenades include:

- **Appearance and materials:**
  - Same shape, weight, and appearance as the live rifle grenade it is designed to resemble.
  - May have an internal or external fuze.
  - Hollow stabilizer tube assembly lacking a venturi or ignition holes.

- **Markings:** A blue body with or without brown markings and a brown band are common. However, black bodies with white markings, yellow bodies with black markings, and other color combinations are also common.

- **Common fuze configurations:** Depending on the design, PD, BD, and PTTF fuzes are direct-armed or partially armed prior to deployment.
- **General safety precautions** for practice rifle grenades vary greatly depending on the model and fuze, but can include:
  - Movement.
  - All applicable safety precautions for the live grenade observed until positive identification is made.
  - Safety precautions for the fuze if present.

Projected Grenades

Projected grenades were devised to cover the tactical range between hand grenades and mortars more efficiently. Projected grenades are configured and constructed as fixed projectiles and meet all three of the defining factors of a projectile provided in Chapter 4:

1. It is fired down a barrel or tube by the gas pressure generated from a propellant charge.
2. The propellant charge is the munition's primary means of deployment.
3. It does not have an attached motor as a primary means of propulsion.

However, projected grenades are defined by tactical design, which is to be deployed by infantry grenadiers. The resulting classification of "ammunition for grenade launchers" and the categorization of projected grenades is a deviation from the conventional category system and something to note for research considerations.

Key Identification Features

Projected grenades usually consist of a body, fuze, energetic material, chemical or inert filler, and a cartridge case if unfired. As a spin-stabilized munition, they also have gas-check bands or rotating bands. Projected grenades are not designed to travel great distances and tend to have an overall lighter appearance. When a projectile with a diameter of 25mm through 40mm is being interrogated, consider the possibility of it being a projected grenade until proven otherwise.

Projected Grenade Sections

Ogive: Projected grenades usually have a flat or rounded ogive. Many 40mm designs have the unique gold anodized ogive seen in Figure 5.37.

Bourrelet: This is the section between the ogive and body, but on many projected grenades the bourrelet is not as obvious as on other projectiles.

Warhead or body: This is the cylindrical section of the projected grenade between the forward bourrelet and the rotating band.

Rotating bands and gas-check bands: As a spin-stabilized munition, projected grenades have rotating or gas-check bands. Gas-check bands are sometimes titled "rotating bands" in reference materials, but they are significantly different in appearance from a conventional copper rotating band (Figure 5.37). There are also designs with a prescored rotating band akin to a recoilless rifle projectile.

Base: There are a number of features associated with the base of a projected grenade that can assist in the identification process. Base configurations can be solid, recessed, contain propellant, or a base fuze that is ignited when fired.

Fuze: A spin-stabilized projected grenade fuze usually requires three actions to arm: (1) **setback** from the acceleration of firing, (2) **centrifugal force** from spin, and (3) **time of flight** during which the actions of (1) and (2) release locks or clockwork mechanisms, initiate pyrotechnic delays, and employ other actions to complete the arming sequence. Nose fuzes are usually observable, but impact with a hard surface may render identification difficult.

The Seven-Step Practical Process Applied to Projected Grenades

Examples of different designs, features, color codes, markings, and construction features are provided throughout this section.

Step 1: Approach and initial interrogation. Attempt to identify a munition at a distance with the use of binoculars. If an approach is made, measurements must be taken of the major diameter at the bourrelet and the overall length of the body from the base to the fuze well. If possible, measure the

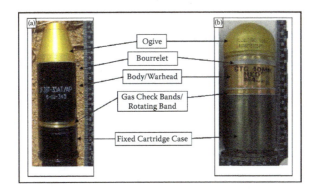

Figure 5.37 (a) Chinese 35mm, DFJ87 AT/AP. (b) U.S. 40mm, M433 HEDP. Both of these munitions are configured with a shaped charge to penetrate armor and a body designed to maximize fragmentation. (Author's photographs.)

location and width of the rotating or gas-check bands. All findings, including measurements, color codes, markings, key identifying features, and any possible damage, are documented and the munition is photographed. In addition to the overall configuration, there are three features that will greatly assist in answering steps 2, 3, 5, and 7:

1. A diameter between 25mm and 40mm.
2. The overall length.
3. The design and material used on the gas-check or rotating bands.

Step 2: Determine fuze type and condition. If a projected grenade has been deployed, the fuze is considered to be armed (step 5). If a fuze is damaged or any alterations have been made to the munition, it is considered armed. If visible, measurements for the fuze are taken separately from the munition.

Step 3: Determine ordnance category. This category covers grenades designed to be fired from a grenade launcher.

Step 4: Determine ordnance group. Identifying characteristics associated with each group will be covered throughout this chapter.

Step 5: Determine if the munition was deployed. Inspect the munition for signs of impact-related damage and the gas-check or rotating bands for scoring.

Step 6: Determine safety precautions that apply to the munition. The safety precautions for the projected grenade groups are covered in this chapter. Chapter 3 addresses the safety precautions associated with various fuzes.

Note: Adhere to all safety precautions that apply.

Step 7: Identify the munition. Identify the projected grenade as well as all possible fuzing configurations.

Groups

The category projected grenade includes these groups:

a. HE (frag).
b. High Explosive Dual Purpose (HEDP).
c. Burning smoke (includes colored smoke and riot control).
d. Illumination.
e. Practice.

3a. Projected grenade, HE (fragmentation): What hand and rifle grenade groups define as "fragmentation" projected grenade literature classifies

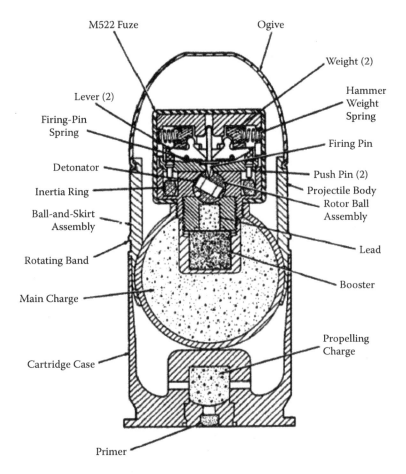

Figure 5.38 Line drawing of a U.S. 40mm, HE, M406 with gas-check bands, which are referred to as "rotating band" in many references. (From U.S. military TM.)

as high explosive. But these are not blast grenades as they usually contain between 1 and 3 oz (28 and 85 g) of high explosives housed in a serrated or microengraved metal body designed to fragment (Figures 5.38–5.40).

General identification features associated with HE projected grenades include:

- **Appearance and materials:**
  - Color codes that deviate from common standards.
  - 25mm to 40mm in diameter.
  - Rotating band or gas-check bands.
  - Aluminum nose fuze housing or rounded aluminum ogive covering the fuze.
  - Usually have a smooth body with internal serrations or a "ribbed" appearance for maximum fragmentation.

Figure 5.39 U.S. 40mm, HE, M406. Note the subdued design of the gas-check bands. (Courtesy of Didzis Jurcins.)

Figure 5.40 Russian 30mm, HE, VOG-17M. With and without cartridge case. (Author's photograph.)

- **Markings:** Color codes used for projected HE grenades are a departure from most standards. Anodized gold-colored ogives are common as are gold, green, black, and yellow body colors with black, yellow, or white markings. Other colors, stamped or stenciled markings, and symbols may also be present.
- **Common fuze configurations:** PD fuzes are common and many designs incorporate a self-destruct (S/D) feature. Variable Time (VT) fuzes are also employed.
- **General safety precautions** for HE projected grenades include:
  - HE, frag, movement.
  - Safety precautions for the fuze if present.

3b. Projected grenade, HEDP (High Explosive Dual Purpose): The United States and some other countries moved away from the single-purpose HE projected grenades in favor of a HEAT configuration housed in a serrated or microengraved metal body designed to penetrate armor and fragment like an HE grenade. For projected grenades, this configuration is defined as High Explosive Dual Purpose (HEDP) or anti-tank/anti-personnel (AT/AP) as seen in Figure 5.37. The internal configuration can be seen in Figure 5.41. With the exception of identifiable markings and perhaps the length of the body, it is usually difficult to distinguish an HE from an HEDP projected grenade.

General identification features associated with HEDP projected grenades include:

- **Appearance and materials:**
  - Color codes that deviate from common standards.
  - 25mm to 40mm in diameter.

Figure 5.41 Far left and right: German 40mm, DM112 HEDP. Center: German 40mm, DM111 HE-PFF (Pre-Formed Fragments). Note the difference in the rotating bands between these grenades and the one shown is Figure 5.39. (Courtesy of Didzis Jurcins.)

- Rotating band or gas-check bands.
- Aluminum nose fuze housing a rounded aluminum ogive covering the fuze.
- A smooth outer body.

- **Markings:** Color codes used for projected HEDP grenades are a departure from most standards. Anodized gold-colored ogives are common, as are gold, green, black, and yellow body colors with black, yellow, or white markings. Other colors, stamped or stenciled markings, and symbols may also be present.
- **Common fuze configurations:** BD or PIBD (spitback) fuzes, with many designs incorporating an S/D feature. VT fuzes may also be employed.
- **General safety precautions** for HEDP projected grenades include:
  - HE, frag, movement, jet.
  - Safety precautions for the fuze if present.

3c. Projected grenade, burning smoke, and riot control: Contain pyrotechnic mixtures that burn to produce smoke in a variety of colors for signaling or screening. The grenade body remains intact during and after smoke production. This configuration is also used to deploy riot-control agents and are also used by law enforcement, usually in 37mm and 38mm diameters. Discussion of riot-control projected grenades will be covered with burning smoke projected grenades versus being afforded their own section as both fundamentally operate the same way (Figure 5.42).

General identification features associated with burning smoke projected grenades include:

- **Appearance and materials:**
  - Color codes that may deviate from common standards.
  - 25mm to 40mm in diameter.
  - Rotating band or gas-check bands.
  - Telltale emission hole or holes in the base.
  - Possible rubber or plastic body for riot-control versions.

- **Markings:** Smoke projected grenades may have a gray or green body with black markings. The ogive color or a band on the body may indicate the color of the smoke produced. Riot-control projected grenades may have a gray or unpainted silver body with red bands or markings. Other colors, stamped or stenciled markings, and symbols may also be present.

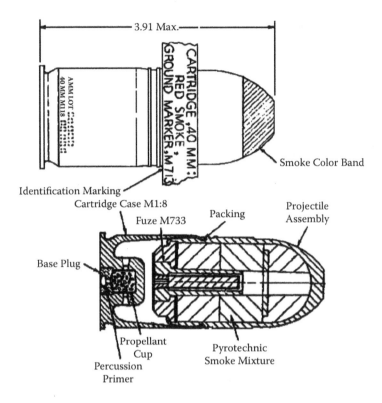

Figure 5.42 Line drawing of a U.S. 40mm, M713 burning smoke. (From U.S. military TM.)

- **Common fuze configurations:** Base fuze with a pyrotechnic delay that is ignited when fired.
- **General safety precautions** for burning smoke projected grenades include:
  - Movement, fire.
  - Chemical if grenade is burning.
  - Safety precautions for the fuze if present.

3d. Projected grenade, Illumination contains pyrotechnic mixtures that burn intensely to illuminate an area or for signaling. Upon firing, a pyrotechnic delay fuze is ignited and functions after a preset time. The fuze functions, and an expelling charge ejects the candle from the grenade body. Depending on the design, a parachute may or may not be present. If a parachute is present, it will deploy and the candle will ignite and burn as it floats toward the ground (Figures 5.43 and 5.44). If a parachute is not present, the candle mixture will burn while falling quickly toward the ground.

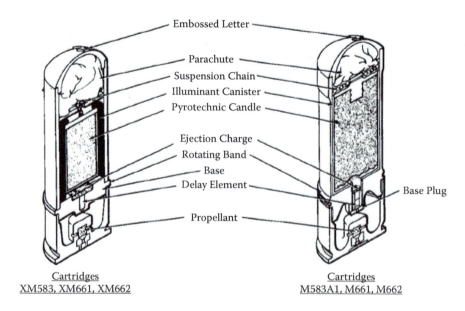

Figure 5.43 Line drawing of the internal configurations of two U.S. 40mm illumination candles with parachutes. (From U.S. military TM.)

Figure 5.44 The ogive of a white illumination candle with parachute. (Author's photograph.)

General identification features associated with illumination projected grenades include:

- **Appearance and materials:**
  - Constructed of aluminum or plastic-coated paper with smooth outer appearance.
  - An internal fuze.
  - Plastic ogive that looks like a cap.

- **Markings:** A white body with black markings is common. Some designs have specific markings. For example, US 40mm illumination projected grenades with parachutes have a slightly dome-shaped ogive embossed with a letter to identify the payload (W = white, G = green, and R = red). On the ogive of star cluster projected grenades there are also five raised dots around the outer edge. Other colors and stamped or stenciled markings, such as the umbrella-like symbols, may also be present.

- **Common fuze configurations:** Two fuzes may be present: (1) a time fuze to eject the payload from the grenade body and ignite the pyrotechnic candle, and (2) depending on the design, the ejection charge may not ignite the candle and a second fuze to perform this task functions after a short delay to ignite the candle.
- **General safety precautions** for illumination rifle grenades include:
  - Movement, ejection, fire.
  - Chemical if candle is burning.
  - Safety precautions for the fuze if present.

3e. Projected grenade, practice: Some practice projected grenades are designed to resemble a live grenade, while others are designed to look and function in the same manner as a live grenade, but lack the full energetic filler of the live munition. "Practice" does not mean safe or inert and, with deployment ranges exceeding 2 km, some practice projected grenades contain live fuzes with substantial spotting charges that allow the point of impact to be seen (Figure 5.45). There are four common configurations of practice projected grenades:

1. Solid one-piece metal construction (on the right in Figure 5.45).
2. A live fuze with an explosive that expels a spotting charge strong enough to lay open the steel base of the grenade (on the left in Figure 5.45).
3. Thin plastic body containing a marking dye that bursts upon impact.
4. A live fuze with smoke pellets in the body that are disbursed upon impact.

General identification features associated with practice projected grenades include:

- **Appearance and materials:**
  - Same shape, weight, and appearance as the live grenade it is designed to resemble.
  - May have an internal or external fuze.
  - Possibly a plastic cartridge case for grenades with a plastic, dye-filled body.
  - Easily mistaken for paintball grenades or children's toys.

Figure 5.45 Right: U.S. 40mm practice grenade consisting of solid aluminum and a copper rotating band. Left: U.S. 40mm M918 with an aluminum ogive, steel body, and copper rotating band. Note the curled steel base where the spotting charge was explosively ejected. (Author's photograph.)

- **Markings:** A blue body with or without brown markings and a brown band is common. However, black bodies with white markings, yellow bodies with black markings, and other color combinations are also common.
- **Common fuze configurations** include a PD fuze if one is present.
- **General safety precautions** vary greatly depending on the model and fuze, but can include:
  - Movement.
  - Observing all applicable safety precautions for the live grenade until positive identification is made.
  - Safety precautions for the fuze and any energetic components if present.

Closing

Grenades are the most commonly encountered ordnance items outside military control. Hand, rifle, and projected grenades are inherently dangerous. Until proven otherwise, always consider a grenade to be in a hazardous condition.

Ordnance Categories—
Aerial Bombs and
Dispensers

6

Viewed from half a world away, a bomb is a political concern; viewed from less than a foot away, a bomb is just a high-stakes exercise in problem solving, where making a mistake means a final, terminal education in the physics of expanding gases.

The Hurt Locker (movie), 2008

Introduction: Aerial Bombs

The quote from the 2008 movie, *The Hurt Locker,* was in reference to Improvised Explosive Devices (IEDs); however, it is also applicable to conventional military ordnance. The largest non-nuclear bomb manufactured by the United States was the T-12 "Cloudmaker," which weighed 43,600 lb (19,800kg) and was classified as an "earthquake" bomb. The largest bomb in the current U.S. inventory is the GBU-43/B Massive Ordnance Airblast Bomb (MOAB) weighing 21,000 lb (9525 kg). The theft or loss of a bomb this size or as small as 500 lb (227 kg) has happened, but is not a common occurrence. The overall focus of this book is the smaller ordnance items more commonly encountered outside military control. This chapter will briefly cover aerial bombs and dispensers.

A basic definition of a bomb is a munition designed to be dropped from an aircraft and fall to the ground. For this chapter, the defining factors that categorize a munition as a "bomb" are:

1. It is dropped from an aircraft.
2. It is not covered by the definitions of rockets or missiles.

Defining factor 1 refers to the means by which bombs are attached to aircraft using one or two "lugs" or "cleats," but two exceptions must be mentioned: (1) Early generations of aerial bombs were slightly modified hand grenades and mortar projectiles dropped from the open cockpit of an aircraft, and (2) large MOAB-type bombs are deployed on a sled that slides out the back of large cargo aircraft. As such, they do not have lugs for exterior attachment to an aircraft.

Defining factor 2 addresses an exception that applies to the guidance systems affixed to "smart bombs" because they provide a bomb with the ability to alter its flight path, which is a defining characteristic of a missile. There are also smart bomb configurations that include a motor on the base fin assembly that propels and extends the bomb's flight path. A motor used for propulsion is a defining characteristic of missiles and rockets, but is an exception associated with bombs.

Delivery Systems

Bombs can be delivered from any airborne platform and the Zeppelins of WWI were initially successful, but vulnerable to ground fire. Helicopters and unmanned drones are capable of carrying bombs, but are much more efficiently armed with missiles and rockets. For over 100 years bombs were and continue to be delivered successfully from fixed-wing aircraft as their primary means of deployment.

Key Identification Features

Most bombs contain a body, a base with a fixed means of orienting the bomb, and a means of attaching the bomb to an aircraft. Though some bombs may contain motors and a means of changing trajectory, these features will be applied to the identification process of rockets and missiles, which are covered in later chapters.

Body designs, which help determine the group to which a bomb belongs, will be covered throughout this chapter. But the key identification features associated with the bomb category remain the configuration of the base with a means of orientation and the lugs required to affix it to an aircraft.

Bomb Sections

Base and orientation: The base of a bomb, including what is attached or fixed to the aft end for stabilization, offers a great deal of information that will assist in identifying an unknown munition. The base and any components affixed to the base are the most likely parts to survive impact and offer evidence to the possible fuzing configuration.

Excluding bombs with transverse fuzing (Figures 3.2, 3.8, and 3.9) in Chapter 3), most bombs contain a fuze well in the base. There are bombs specifically designed without orientation, which will be covered in this chapter, under group 2c. However, most bombs contain fins that fall into two categories, low and high drag:

- **High-drag fins:** Are designed to slow a bomb as it descends, allowing fast moving aircraft to deliver bombs at low levels without being damaged by the bombs they drop. Examples of fins designed for retardation are parachutes housed in large, square-shaped fin assemblies

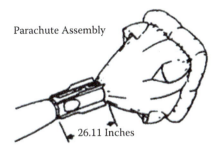

Parachute Assembly

26.11 Inches

Figure 6.1 Line drawing of a U.S. parachute retardation fin assembly. (From U.S. military TM.)

(Figure 6.1) and "Snakeye" fins (Figure 6.2). The violent opening shock of the parachute or Snakeye fin can be part of the arming sequence for bomb fuzes designed to be deployed for this tactical purpose.

- **Low-drag fins:** Aerodynamically orient a bomb while offering no drag. Designs include the conventional four-fin conical, the four-fin boxed style (Figure 6.3), and a multi-fin enclosed configuration (Figure 6.4).

Figure 6.2 U.S. Snakeye retardation fin assembly mounted on an MK-82, 500 lb (226 kg) bomb. The green body with the unconventional white marking designates an inert filler. (Author's photograph.)

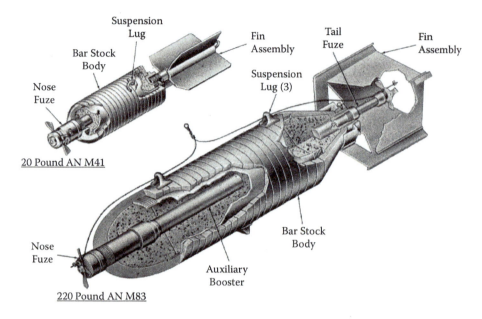

Figure 6.3 Conventional low-drag fin designs on a fragmentation bomb (top) and a General Purpose (GP) old-style bomb design. (From U.S. military TM.)

Figure 6.4 A low-drag "captured" fin design commonly associated with bombs from China, Russia, and other Eastern European countries. (Author's photograph.)

Figure 6.5 An old-style design with Bulgarian markings. Note the two-lug and one-lug configurations 180° on the bomb body. (Author's photograph.)

Attachment to aircraft: Excluding bombs dropped from an open cockpit or out of the back of large cargo aircraft, lugs or "cleats" are the most common means of attaching a bomb to an aircraft. Most aircraft employ a two-lug configuration, but some use a single lug. Bombs that can be dropped from various aircraft types often have two lugs on the side of the body with a single lug on the body 180° from the two lugs (Figures 6.3 and 6.5).

If a bomb were simply dropped, the aerodynamics of many aircraft would cause the bomb to repeatedly strike and damage the aircraft. In order to negate this possibility, a "crow's foot" powered by an explosive cartridge housed in the ejection rack is positioned in contact with the bomb, between the lugs. When the bomb is released the crow's foot is explosively thrust outward approximately 1 ft, jettisoning the bomb away from the aircraft.

Fuzing: Bomb fuzes can be armed a number of ways. Electrical fuzes may have arming wires pulled and receive a surge of power from the aircraft as it is released, or an onboard wind generator may spin to provide power as the bomb falls. Mechanical fuzes may have arming wires pulled as they are jettisoned from the aircraft, freeing other components to move and arm the fuze as the bomb falls.

The Seven-Step Practical Process Applied to Bombs

Examples of different designs, features, color codes, markings, and construction features are provided throughout this section.

Step 1: Approach and initial interrogation. Attempt to identify a munition at a distance with the use of binoculars. If an approach is made, avoid all venturis and fuze-sensing elements. Armed and active or damaged sensing elements may "see" a person approaching, consider the person a valid target, and function as designed.

At a minimum, measurements must be taken of the major diameter at midbody, the length of the munition, and the distance between the lugs. Look for stamped data on the munition, focusing on the area between the lugs and on the base. Note the texture of the body; is it smooth or rough with a "gator-skin" like coating? All findings, including measurements, color codes, markings, key identifying features, and any possible damage, are documented and the munition is photographed.

In addition to the overall configuration, there are four features that will greatly assist in answering steps 2, 3, 5, and 7:

1. The diameter.
2. The overall length.
3. The method of orientation (i.e., high or low drag).
4. Whether the body is one-piece steel construction or multiple-piece construction of light materials.

Step 2: Determine fuze type and condition. If a bomb has been deployed, the fuze is considered to be armed (step 5). If a fuze is damaged, pins have been removed, or any other alterations have been made to the munition, it is considered armed. If visible, measurements for the fuze are taken separately from the munition.

Step 3: Determine ordnance category. This category covers bombs designed to be dropped from an aircraft.

Step 4: Determine ordnance group. Identifying characteristics associated with each bomb group will be covered throughout this chapter.

Step 5: Determine if the munition was deployed. Inspect the bomb for impact-related damage and missing pins or clips from the fuze.

Step 6: Determine safety precautions that apply to the munition. The relevant safety precautions for bombs are covered in this chapter. Chapter 3 addresses the safety precautions associated with various fuzes.

Note: Adhere to all safety precautions that apply to the bomb and fuzing.

Step 7: Identify the munition. Apply the totality of all construction characteristics and other identifying features to determine the group to which a bomb belongs and, if feasible, positively identify the munition and all possible fuzing configurations.

Groups

In order to provide a coherent flow, the category bomb is divided into the following primary and supplemental groups:

1. High Explosive.
   a. Fragmentation.
   b. General Purpose (GP) old style.
   c. GP new style.
   d. GP demolition bomb.

     e. Penetration.
     f. Guided.
     g. Fuel Air Explosive (FAE).

2. Fire:
     a. Photoflash.
     b. White Phosphorus (WP).
     c. Napalm.

3. Practice:
     a. With explosives or spotting charges.
     b. Inert.

1. Bomb, HE: The bombs covered under this group are designed to explode and produce destructive effects through blast pressures and fragmentation. However, if deployed properly, penetration bombs will detonate subsurface, thus negating fragmentation effects, and (1e) produce only blast pressure effects. Explosive fillers in these bombs range from a few pounds to hundreds of pounds and may be solid, pliable, or, for FAE bombs (1g), in liquid form. Each contains different construction design features that may allow positive identification and the application of appropriate safety precautions. Six of the seven high-explosive (HE) type bombs are closely related. Only FAE bombs (1g) operate in a significantly different way.

The examples provided in this section do not include bombs with transverse fuzing, which are shown in chapter 3, "Fuze Functioning."

1a. Fragmentation bombs: Contain high explosives housed in a metal body with additional internal or external fragmentation. Figure 6.3 provides two examples. Lugs are usually welded to the body.

1b. GP old-style bombs: Contain high explosives housed in a metal body without additional fragmentation. The bomb in the bottom of Figure 6.3 (without the added fragmentation sleeve) and that in Figure 6.5 are examples of different body configurations. Most GP old-style designs are for aircraft with internal bomb bays as the shape may cause drag-related issues. Lugs are usually welded to the body.

1c. GP new-style bombs: Contain high explosives housed in a metal body without additional fragmentation. GP new-style designs are sleek for external carry and contain internal electrical plumbing to allow various electric fuzes to be used (Figures 6.6, 6.7, and 6.8). Lugs may be threaded into the body and can be removed to roll the bomb.

1d. GP demolition bombs: Contain high explosives housed in a metal body without additional fragmentation. The designation "demolition bomb" is applied to bombs with thin bodies designed to carry a heavier explosive charge and produce greater blast effects than that of GP bombs. The body

Figure 6.6 Egyptian 100 kg (220 lb) GP bomb with single welded lug and a parachute retardation fin assembly. (Author's photograph.)

will fragment—but with an emphasis on producing damage with the blast; fragmentation effects are greatly reduced. Lugs are usually threaded into the body and can be removed to roll the bomb.

1e. Penetration bombs contain high explosives housed in a sleek, thick, hardened steel body with a hardened-pointed nose. These bombs are designed to use kinetic energy generated during their descent to penetrate deeply into the ground or other hardened targets before being detonated by a base fuze. Upon detonation, the blast effects result in overpressures and shockwaves traveling through the ground that are capable of collapsing underground facilities and infrastructure. Though penetration bombs have a thick body design capable of producing significant fragmentation, this effect is negated by the environment in which they are intended to function (Figure 6.7). Penetration bombs can also be fitted with guidance systems to increase accuracy. Lugs are usually threaded into the body and can be removed to roll the bomb.

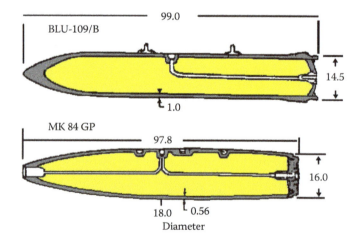

Figure 6.7 U.S. BLU-109/B penetration bomb compared to a 2000 lb (906 kg) MK-84 GP new-style bomb. (From U.S. military TM.)

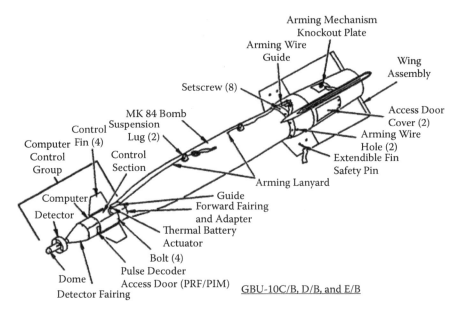

Figure 6.8 Line drawing of an older model guided bomb unit (GBU)-10. Note the four control fins on the forward guidance section, which steer the bomb to its target. (From U.S. military TM.)

1f. Guided bombs: Are a GP new-style, or penetration bombs fitted with a guidance section and steerable fins on the nose and oversized fins on the base. The guidance section has the ability to manipulate the fins so as to steer the bomb toward its intended target. This designation covers systems made by all countries, including the older Guided Bomb Unit (GBU) and newer Joint Direct Attack Munition (JDAM) employed by the United States (Figure 6.8).

1g. Fuel Air Explosive (FAE): Bombs contain a high-explosive burster housed in a thin outer body filled with a fuel-rich material. When functioned, the body will fragment as the burster aerosolizes the dense fuel filler and deploys cloud detonators. As the aerosolized fuel mixes with air, as a stoichiometric ratio is reached, the cloud detonators function, resulting in an explosion with extremely strong blast effects. The metal body may be prestressed to fragment easily into large, thin pieces that do not travel far. Additional information is provided in Chapter 9, group 1c, and a cutaway of an FAE bomb is provided in Figure 8 of that chapter.

General information associated with high-explosive bombs includes:

- **Markings for explosive bombs:** A gray, green, black, or brown body with yellow, red, or black markings is common. Stamped or stenciled markings, or a data plate between the lugs and/or on the base-plate is common. Other colors, stamped or stenciled markings, and symbols may be present.

- **Common fuze configurations:** Point Detonating (PD), Base
  Detonating (BD), Point Initiating Base Detonating (PIBD), time
  fuzes, Variable Time (VT), pressure, and influence fuzing configura-
  tions are all used with bombs. Depending on the design, fuzes can be
  located in the nose, tail, or side (transverse) of the bomb.
- **General safety precautions** for high-explosive bombs include:
  - HE, fragmentation (frag), movement.
  - Safety precautions for the fuze if present.

Note: For guidance systems add electromagnetic radiation (EMR), static,
and, possibly, ejection for the fins.

Note: FAE mixtures vary and many are toxic, so, if suspected, add chemical.

2. Bomb, fire: Although the fuel sources and temperatures vary, all
three bomb designs covered under this group produce fire; two also generate
extreme heat.

2a. Photoflash bombs: Contain a photoflash mixture (Chapter 1)
that burns with explosive force and produces an intense flash of light and
extremely high temperatures. The United States deployed these bombs with
time fuzes so they would function above the ground. Capable of producing
over a billion candlepower, photoflash bombs were used to support reconnaissance photography by airbursting as a giant flashbulb. A thin body of
lightweight material, as well as the strap-lug configuration in Figure 6.9, is
often used with photoflash bombs. As most of these bombs are older designs,
box fins are common (Figures 6.3 and 6.9). Though obsolete, an old photoflash bomb with a deteriorating case offers a unique hazard. Photoflash powder reacts with water, resulting in the generation of heat and the production
of hydrogen gas, which may initiate the munition.

2b. White Phosphorus (WP) bombs: Contain WP sealed in a thin body of
lightweight material with a high-explosive burster running through the bomb's
center. When functioned, the burster detonates, breaking the bomb body into
pieces while dispersing the WP. The strap-lug configuration in Figure 6.9 and
the fin designs in Figure 6.3 are common on these obsolete bombs.

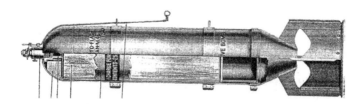

Figure 6.9 Line drawing of a U.S. M46 photoflash bomb, which is capable of
producing 500,000,000 candlepower. (From U.S. military TM.)

General information associated with photoflash and WP bombs includes:

- **Markings for photoflash and WP bombs:** A gray or lime-green body with yellow, red, or black markings is common. Other colors, stamped or stenciled markings, and symbols may be present.
- **Common fuze configurations:** Mechanical Time (MT), Powder Train Time Fuze (PTTF) with photoflash bombs, and PD fuzes with WP bombs. Depending on the design, fuzes can be located in the nose, tail, or side of the bomb.
- **General safety precautions** for photoflash and WP bombs include:
  - HE, frag, movement, fire, chemical, WP.
  - Safety precautions for the fuze if present.

Note: If burning, WP smoke is toxic.

2c. Fire bomb—napalm: Consists of a fuel and a thickening agent mixed to produce a jelly-like material contained in a thin aluminum body or container. The unique aspects that differentiate this from other fire-producing bombs are the design of the body and functioning sequence. A napalm container is designed to be out of balance and is deployed as a fin-less munition to ensure that it tumbles while falling to the ground. Upon impact, the thin body tears open, dispersing napalm as the fuzes function, resulting in a widespread fireball. Threaded lugs and numerous fuze wells in the ends and side are also common configurations of napalm-filled fire bombs (Figure 6.10).

General information associated with napalm-filled bombs includes:

- **Markings for napalm-filled bombs:** An aluminum body that may be unpainted with red or black markings is common. Other colors, stamped or stenciled markings, and symbols may be present.
- **Common fuze configurations:** The lack of orientation prior to impact and the fact that napalm-filled fire bombs are designed to burst open upon impact requires a unique fuze and booster configuration. An all-way-acting fuze is employed as, regardless of the bomb's orientation upon impact, the fuze will function (Figure 3.14 in Chapter 3). To ensure that napalm spread by the violent impact is ignited, a WP or Magnesium Teflon igniter (booster) is used. Due to the complexity of this deployment configuration, two fuzes are commonly employed. For example, the MK77 in Figure 6.10 contains four fuze/igniter wells: one on each end and two transverse fuze wells.
- **General safety precautions** for napalm-filled bombs include:
  - HE, frag, movement, fire.
  - Safety precautions for the fuze if present.

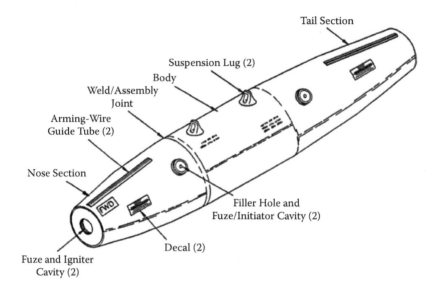

Figure 6.10 Line drawing of a U.S. MK77 firebomb. (From U.S. military TM.)

3. Bomb, practice, with explosives or spotting charges: Are designed to be dropped and fall with the same ballistic characteristics as an explosive-filled bomb. Some practice bomb configurations include a solid body, empty body, or inert filler, but many include live fuzes, substantial explosive charges, or spotting charges that are explosively ejected. Spotting charges also offer unique hazards, including RP and glass vials containing titaniumtetracloride that are explosively ejected upon fuze functioning.

General identification features associated with practice bombs include (Figure 6.11):

- **Appearance and materials:** The construction features are associated with the bomb they are designed to imitate.
- **Markings:** A blue, black, or green body with white markings or brown bands is common. Other colors, stamped or stenciled markings, and symbols may also be present.
- **Common fuze configurations** depend on bomb type.
- **General safety precautions** for practice bombs include:
  - Movement.
  - Observing all applicable safety precautions for the live bomb until positive identification is made.
  - HE, frag, ejection when a spotting charge is present.
  - Safety precautions for the fuze if present.

Figure 6.11 From top to bottom: Bomb Dummy Unit (BDU) 33, 25 lb practice bomb; MK 106, 5 lb practice bomb; MK 3, 3 lb practice bomb. Example of a signal cartridge that can be used with all three; which may contain Red Phosphorus (RP) or titaniumtetracloride. (Author's photograph.)

Note: Fire, if a RP spotting charge is suspected.
Note: Chemical, if a titaniumtetracloride spotting charge is suspected.
Note: Practice means "practice"—not "inert."

Introduction: Dispensers

Aerial dispensers deploy submunitions or mines from aircraft in two different ways, which define their categories. For aerial dispensers, the category is dispenser and the two groups are:

1. Retained.
2. Dropped.

   A dispenser is positioned on an aircraft in the same configuration as a bomb. The primary difference between retained and dropped dispensers is that a retained dispenser remains fixed to the aircraft when its payload is deployed, while a dropped dispenser is deployed from the aircraft in a similar fashion as a bomb. Dispensers can be filled with a variety of payloads, including submunitions, mines, and other miscellaneous items such as leaflets. The focus of this chapter will be on dispensers filled with submunitions and mines.

Key Identification Features

Dispensers are fixed with lugs or cleats to secure them to an aircraft. These munitions are characterized by a thin-skin nonaerodynamic body shape that oftentimes includes unique markings.

The Seven-Step Practical Process Applied to Aerial Groups

Examples of different designs, features, color codes, markings, and construction features are provided throughout this section.

Step 1: Approach and initial interrogation. Attempt to identify a munition at a distance with the use of binoculars. If an approach is made, avoid all venturis and fuze-sensing elements. Armed and active or damaged sensing elements may "see" a person approaching, consider the person a valid target, and function as designed.

At a minimum, measurements must be taken of the major diameter at the midbody—between the lugs as well as the distance between the lugs—and the overall length. Look for stamped and stenciled data on the munition, focusing on the areas between the lugs and on the base. All findings, including measurements, color codes, markings, key identifying features, and any possible damage are documented and the munition is photographed.

In addition to the overall configuration, there are four features that will greatly assist in answering steps 2, 3, 5, and 7:

3. The diameter.
4. The overall length.
5. Holes or covered round sections on any side of the munition.
6. Seams running the length of the body and other evidence of multiple-piece construction.

Step 2: Determine fuze type and condition. Retained dispensers do not have a "fuze" as defined in this text, but do contain sequencers, initiators, and other fuze-like components to deploy their contents.

If a dispenser has been deployed, the fuze is considered to be armed (step 5). If a fuze is damaged, pins removed, or any alterations made to the munition, it is considered armed. If visible, measurements for the fuze are taken separately from the munition.

Step 3: Determine ordnance category. This category covers both retained and deployed dispensers.

Step 4: Determine ordnance group. Identifying characteristics associated with retained and deployed dispensers will be covered in this section.

Step 5: Determine if the munition was deployed. Inspect the dispenser for impact-related damage and missing pins or clips from the fuze.

Step 6: Determine safety precautions that apply to the munition. The relevant safety precautions are covered in this section. Chapter 3 addresses safety precautions associated with various fuzes.

Note: Adhere to all safety precautions that apply to the dispenser, its payload, and fuzing.

Step 7: Identify the munition. Apply the totality of all construction characteristics and other identifying features to determine the group to which a dispenser belongs and, if feasible, positively identify the dispenser and all possible payload and fuzing configurations.

1. Retained dispensers: Are used to deploy payloads in a number of ways including unlocking the payload, allowing it to free-fall, or forcible deployment with ram-air produced by the speed of the aircraft. Springs and smokeless powder-filled Cartridge Actuated Devices (CADs) are also used to eject munitions forcibly from a dispenser. Retained dispensers are not damaged during payload deployment. They are designed to be refilled and used numerous times during their service life span.

General identification features associated with a retained dispenser include (Figure 6.12):

- **Appearance and materials:**
  - Multi-piece, thin skin aluminum or tin body.
  - No fuze.
  - A canoe or bounded tube-like configuration that allows payloads to be jettisoned from the rear or bottom of the dispenser.

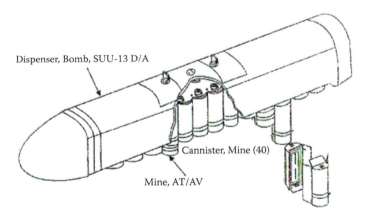

Figure 6.12 Line drawing of an SUU-13 retained dispenser capable of deploying a variety of mines, submunitions, and tactical smoke munitions. The configuration shown has the SUU-13 loaded with M56 AT mines. (From U.S. military TM.)

- **Markings:** Stamped or stenciled markings and symbols are usually present.
- **Common fuze configurations:** A retained dispenser cannot deploy the entirety of its payload at one time as the contents will be damaged impacting each other. To avoid this issue, retained dispensers usually incorporate timing sequencers or other means of deploying single submunitions in quick succession.
- **General safety precautions** for an aerial retained dispenser filled with a payload include:
  - Movement, ejection.
  - Safety precautions for any fuzing that may be present.

Note: Depending on the payload, HE, frag, jet, WP, fire, and chemical may apply.
Note: Depending on the dispenser, EMR and static may apply.
Note: The payload may be energetically deployed and offers a substantial ejection hazard.

2. Dropped dispenser: Deployed in the same manner as a bomb, dropped dispensers have fixed, low-drag fins for stabilization on the aft end and a nose fuze, because base and transverse fuzing are generally not compatible with these munitions. Dropped dispensers come in a variety of shapes depending on the payload, but most do not have an aerodynamic configuration. The payload may consist of 3 to 3,000 submunitions or mines that are deployed when the fuze functions at a predetermined height above the target. Depending on the design, fuze functioning may initiate linear shaped charges that explosively open the dispenser lengthwise, or detonating cord that blows off the base to deploy the payload. Other dispensers deploy their contents when locks securing the outer body are released; others explosively inflate bladders to propel the contents outward.

General identification features associated with a dropped dispenser include (Figure 6.13):

- **Appearance and materials:**
  - Multi-piece, with thin skin aluminum or tin body.
  - Fins may be spring loaded and held in place by a band that is released when dropped.
  - There is a basic lack of aerodynamic shape normally associated with aviation ordnance.

- **Markings:** The payload will define the color codes and markings. Stamped or stenciled markings and special symbols may also be present.
- **Common fuze configurations:** MT, PTTF, and Electronic Time (ET) fuzing.

Figure 6.13 A CBU-52B/B dropped dispenser. Note the two-lug configuration, riveted fins, and noticeable break down the length of the body. (Author's photograph.)

- **General safety precautions** for an aerial dropped dispenser filled with a payload include:
  - Movement, ejection.
  - Safety precautions for the fuze if present.

Note: Depending on the payload, HE, frag, jet, WP, fire, and chemical may apply.

Note: Depending on the dispenser, EMR and static may apply.

Note: The payload and fins may be energetically deployed and offer substantial ejection hazards.

Closing

Large bombs and dispensers are not commonly encountered outside of military control, but smaller practice bombs are and they pose a significant hazard. Until proven otherwise, always consider a bomb, including smaller practice models, to be in a hazardous condition.

Ordnance Category—Rockets

7

I certainly remember building model rockets. It was fun to watch the rocket blast into the air, suspenseful to wonder if the parachute would open to bring the rocket safely back.

Eric Allin Cornell, winner of 2001 Nobel Prize in physics

Introduction

The history of rocket development is contested by many historians, but may date back as far as AD 300. The first documented use of rockets in combat dates to the eleventh century. Most Americans associate rockets with a line from "The Star-Spangled Banner": "...the rocket's red glare, the bombs bursting in air...," which refers to British Congreve rockets fired at Fort McHenry from ships in Baltimore Harbor.

Early rocket designs used black powder as a propellant and as the warhead's main charge. Sir William Congreve successfully designed several different warheads that were fired with a long, wooden shaft intended to provide stabilization. The next leap in "rocket science" was achieved by British inventor William Hale, who successfully incorporated angled venturis in the base of a rocket motor to generate gyroscopic stability, or spin stabilization, which greatly improved accuracy and allowed the wooden shaft to be removed. Still, with black powder as the primary propellant, there were many incidents of rockets exploding in place when fired. These malfunctions resulted in a lack of trust in rocket technology, which limited the use of rockets for the next 70 years.

During WWI, American physicist Dr. Robert Goddard had great success when he replaced black powder with double-base powder, which led to further developments in propellant mixtures. These seemingly small changes allowed fielding of lighter rockets that could be carried by infantrymen. A new concept that came to be known as a "bazooka" is still applied to the shoulder-fired rockets of today.

Of all the ordnance categories, rockets have always seemed to fascinate people more than any other ordnance category, as evidenced by Dr. Cornell's quote. Despite all the accomplishments this man has achieved, he still vividly recalls the enjoyment of a launch and the suspense associated with the intended parachute deployment from a child's hobby rocket. The fascination with a launch may also explain why modern day fireworks shows are so popular.

For this chapter, the defining factors that categorize a munition as a "rocket" is that the body being projected:

1. Is propelled by a rocket motor. This definition is easier to recognize if the motor remains attached to the warhead during flight, but this is not always true.
2. Is unguided and incapable of altering its trajectory while in flight.

There are exceptions to this definition, including terminology such as "RPG" (Rocket-Propelled Grenade) used to describe a reloadable infantry weapon. The term RPG represents an attempt at a literal translation that was incorrect, but generally accepted by the international community. There are versions of these munitions that are categorized as rockets or projectiles, depending upon how they are propelled.

Rocket Types

Many rocket designs can be deployed from more than one platform. But the first defining characteristic of a rocket is its "type," which is a classification derived from the launch point. Modern rockets are fired from **surface**-to-surface or from **air**-to-surface. If the rocket type can be ascertained, it will help in the identification process.

Surface rockets: Are fired from the ground or water by infantrymen, vehicles, unmanned mounts, and ships. These are referred to as surface-to-surface rockets. Due to the ability of vehicles to carry extremely heavy loads, surface-fired rockets can be very large.

Air rockets: Are fired from aircraft, both fixed-wing aircraft and helicopters, including unmanned drones. These are referred to as air-to-surface rockets. Weight limits associated with aircraft define the size of the rockets they can carry.

Key Identification Features

A rocket is a long, cylindrically shaped munition with a warhead and one or more motors containing one or more venturis and no evidence of steerable fins that would allow the munition to change direction in flight. One consideration is that rocket warheads share many shape consistencies with projectiles of the same group. If a rocket warhead is detached from the motor, it is easily mistaken for a projectile. In such a case there are two areas to inspect: (1) The base of a rocket warhead will have a means of affixing to a motor, such as a threaded or grooved base that screws or clips into the motor, and (2) rockets lack the rotating or gas-check band(s) consistent with a projectile (Figure 7.1).

Figure 7.1 2.75 in. rocket warheads. The blue warhead on top clips onto the motor; the black warhead on the bottom is screwed onto the motor. (Author's photograph.)

Rocket Sections

The definitions of ogive, bourrelet, warhead, and body provided in Chapter 4 also apply to rockets.

Base or motor section: For clarification purposes, the base of a rocket is the aft end of the motor (Figures 7.2–7.4) or the fin assembly as some motors are consumed when fired (Figures 7.5–7.7). The bases of many rockets have exposed igniter lead wires, shorting clips, or shunting

Figure 7.2 At left, 107mm and, right, 240mm spin-stabilized rockets. Note that the venturis are angled to impart spin. (Author's photograph.)

Figure 7.3 A fin-stabilized rocket with fixed fins and a single straight venturi. (Author's photograph.)

Figure 7.4 A fin-stabilized rocket with fixed fins, straight venturi, and observable lead wires. (Author's photograph.)

Figure 7.5 RPG launcher, with PG-7 rocket. (Author's photograph.)

Figure 7.6 Prior to firing, PG-7 rocket fins folded within the propellant. (Author's photograph.)

covers that can assist with identification, but they are easily damaged (Figure 7.4).

Motors: Rocket motors consist of a body, igniter, propellant, and one or more venturis. When fired, the igniter initiates the propellant, which generates pressure within the body, which is controlled by the venturi design. Depending on the design, the base or aft end of the motor will often have one or more venturis (Figures 7.2 and 7.3). Some rocket designs include a two-motor configuration with a launch motor and flight motor to increase range and stability. For example, the RPG-7 launch motor (Figures 7.5 and 7.6) fires the rocket out of the weapon system, allows the fins to deploy, and initiates the fuze arming process; then the flight motor fires (Figure 7.8) which keeps the rocket on its intended trajectory to complete the fuze-arming process.

Venturi or nozzle assembly: The diameter, pitch, slant angle, and canting or fluting of venturis, as well as the number of venturis, control the release and direction of pressure generated by the propellant. How this

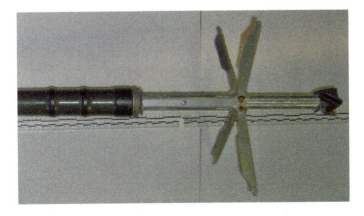

Figure 7.7 After firing, PG-7 fins deployed. (Author's photograph.)

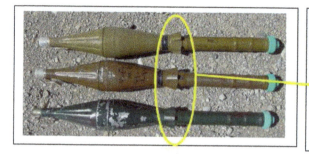

(a)

Figure 7.8 The venturis are at the front of the flight motor on these PG rockets from three different countries. The number of venturis may help identify the country of manufacture. (Author's photograph.)

is done greatly influences the intended trajectory of the munition. When inspecting a rocket, avoid the areas in front of the munition as well as behind the venturis. To determine if a motor has been fired, use a mirror to inspect the venturi(s) and establish if the closure disk is missing. In Figure 7.2, the closure disks are blown out and not present; in Figures 7.3 and 7.8(a), the closure disks are in place. Design details of the venturi(s) greatly assist in identifying a rocket; always count the number of venturis, measure their diameter, and note the venturi type(s) (i.e., straight, slanted, canted, or fluted).

Fin assemblies: Fins are not always present as rockets can be fin- or spin-stabilized. When present, fins may be mounted on the motor or on a separate component of the munition. They may be fixed and immovable (Figure 7.3) or open after deployment (Figure 7.7). If present, fins offer a differentiating identification feature between a rocket and a missile. On a rocket, all the fins are fixed upon deployment, or fixed after opening. Missiles have a means of changing direction, which is usually accomplished with movable fins, a feature rockets should not possess.

Fuze: Many rocket fuzes require three actions to arm and these actions coincide with the way in which a rocket is deployed—for example:

1. A spin-stabilized rocket fuze usually requires three actions: (1) **setback** from the acceleration of firing, (2) **centrifugal force** from spin, and (3) **time of flight** during which the actions of (1) and (2) release locks or clockwork mechanisms and initiate pyrotechnic delays and other actions to complete the arming sequence.
2. A fin-stabilized rocket that does not spin may employ a fuze with (1) **pins or clips** that must be removed prior to firing, and then (2) **setback** from the acceleration of firing, and (3) **time of flight** during

which the actions of (1) and (2) release locks or clockwork mechanisms and initiate pyrotechnic delays and other actions to complete
the arming sequence.

3. A rocket with two motors may employ a fuze that requires (1) **setback** from the acceleration of firing, (2) **deceleration** as the munition slows during flight, and (3) **setback** when the second motor fires,
releasing locks or clockwork mechanisms, initiating pyrotechnic
delays, and other actions to complete the arming sequence.

Rocket nose fuzes are usually observable, but impact with a hard surface may render identification difficult. If a nose fuze is sheared off flush
with the fuze well, the components required to function the fuze may still
be present. Base fuzes are covered by the motor and will not be observable
unless the warhead was removed from the motor prior to discovery. Many
warhead configurations will provide insight to the presence of a fuze, but
positive identification is required to determine conclusively if a fuze is
present.

If a nose or base fuze can be seen, the wrench flats, spanner holes or slots,
and the overall construction will provide relevant information to its identity.
Additionally, the presence of a fuze adapter or booster adapter between the
fuze and the projectile may help identify both the fuze and the rocket.

There are also rocket fuzes that are internal to the munition and offer
few clues to their existence. An example would be the BD element of a PIBD
fuze.

Warhead: The warhead defines the group to which a rocket belongs.
Rocket warheads share many characteristics found on projectiles, with a
base capable of attaching to a motor and the absence of rotating or gas-check
bands, all of which are absent in Figure 7.1.

Warning: If a rocket warhead is found with its motor broken off, do not
assume the base of the warhead is observable. Rocket motors may break off
below the "cup" into or onto which the warhead is screwed or snapped, and
this piece is covering the base of the warhead.

The Seven-Step Practical Process Applied to Rockets

Examples of different designs, color codes, markings, and construction features are provided throughout this chapter.

Step 1: Approach and initial interrogation. Attempt to identify a munition at a distance with the use of binoculars. If an approach is made, avoid all
venturis and fuze-sensing elements. Armed and active or damaged sensing
elements may "see" a person approaching, consider the person a valid target,

and function as designed. At a minimum, measurements must be taken of the major diameter at the bourrelet and the overall length of the body from the base to the fuze well. Separate measurements of the warhead and motor must be taken from where they meet to the base of the motor and to the forward end of the warhead, or the fuze well if a nose fuze is present. Look for stamped data on the munition and, if present, focus on the area just above the warhead–motor junction. All findings, including measurements, color codes, markings, key identifying features, and any possible damage, are documented and the munition is photographed.

In addition to the overall configuration, there are four features that will greatly assist in answering steps 2, 3, 5, and 7:

1. The diameter.
2. The overall length.
3. The method of stabilization (i.e., spin, fin, or a combination of both).
4. Whether the motor is fired or unfired.

Step 2: Determine fuze type and condition. If a rocket has been deployed, the fuze is considered to be armed (step 5). If a fuze is damaged, pins have been removed, or any alterations have been made to the munition, it is considered armed. If visible, measurements for the fuze are taken separately from the munition.

Step 3: Determine ordnance category. This category covers rockets, which are fired by a propellant-filled motor as the primary means of deployment. A brown-colored band may indicate the presence of a rocket motor.

Step 4: Determine ordnance group. Identifying characteristics associated with each rocket group will be covered throughout this chapter; many are consistent with the grouping characteristics of other categories.

Step 5: Determine if the munition was deployed. Inspect the rocket warhead for impact-related damage. Inspect the venturi(s) for the presence of a closure disk or plug. If a closure disk is present, the motor has not been fired. If closure disks are missing, assume the motor has been fired and the fuzing has armed (step 2).

Step 6: Determine safety precautions that apply to the munition. The safety precautions for the rocket groups are covered in this chapter. Chapter 2 addresses the safety precautions associated with various fuzes. Additionally, electromagnetic radiation (EMR), static, and ejection apply to all unfired rocket motors.

Note: Adhere to all safety precautions that may apply to the rocket, motor, and possible fuzing.

Step 7: Identify the munition. Apply the totality of all construction characteristics and other identifying features to determine the group to which a rocket belongs and, if feasible, positively identify the munition and all possible fuzing configurations.

Groups

Per the definition provided in the introduction of this chapter, the rocket category encompasses thousands of different ordnance items. In order to provide a coherent flow, the rocket category is divided into the following primary and supplemental groups:

1. High Explosive (HE).
   a. HE/fragmentation (frag).
   b. Bounding.
   c. Thermobaric.

2. HEAT.
3. Dispenser.
4. Bursting smoke.
5. Illumination.
6. Practice.
   a. With and without spotting charges.
   b. Drill and dummy.

1. Rocket, high explosive: Are designed to explode and produce destructive effects through blast pressures and/or fragmentation. Explosive fillers in these rockets range from a few ounces to hundreds of pounds and may be in solid or liquid form. Construction design features provide evidence that assists in the identification process, with the objective of applying the appropriate safety precautions.

1a. Rocket, HE/frag: Is a general purpose rocket configuration that can be used on a number of different target types. HE rockets usually consist of a motor, warhead, explosive filler, and fuzing. Rocket warheads with enhanced fragmentation effects are grouped under HE/frag. Additionally, there are Semi-Armor-Piercing High-Explosive (SAPHE) warheads also grouped under HE/frag, but these are not common or as effective as APHE projectiles.

General identification features associated with HE/frag rockets include (Figures 7.9–7.11):

- **Appearance and materials:**
  - Warhead, a one-piece body of robust construction.
  - The United States does not commonly use fuze adapters with HE ordnance, but other countries do. When present, top-down slots or side spanner holes on a fuze adapter may indicate if the warhead is filled with HE or WP.

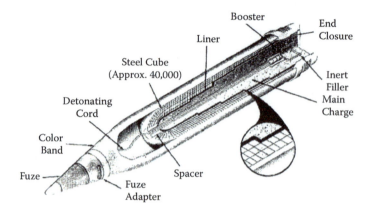

Figure 7.9 Line drawing of an enhanced fragmentation warhead with a PIBD, VT fuze to ensure an air-burst, focusing the majority of fragmentation front-ward. This is an obsolete design. (From U.S. military TM)

- HE/frag rockets have a motor on the aft end with the warhead on the forward end.
- Fin assembly on the aft end of the motor.
- One or more venturis on the base of the motor.

- **Markings:** A green or gray body with yellow or black markings is a common warhead color code. A brown band may be present on the motor. Other colors, stamped or stenciled markings, and symbols may also be present.
- **Common fuze configurations:** Point Detonating (PD), Base Detonating (BD), PIBD, and Variable Time (VT) fuzing. Many war-head designs include a nose and base fuze, or a base fuze only that is covered by the motor.
- **General safety precautions** for HE/frag rockets include:
  - HE, frag, movement, and ejection; EMR, static, and ejection for an unfired motor.
  - Safety precautions for the fuze if present.

Figure 7.10 An OG-7V HE rocket, which is fired from an RPG. The green cap at the base is removed in order to attach the propellant–fin assembly seen in Figures 7.6 and 7.7. (Author's photograph.)

Figure 7.11 Cutaway view of a basic HE/frag rocket with a point-detonating (PD) fuze. (Author's photograph.)

1b. Rocket, HE bounding: There are not many versions of this subgroup, but the Chinese Type 69 in Figure 7.12 is the most common rocket of this design. The Type 69 fuze will function upon direct impact with a target or upon a glancing impact when the "scoop" or "collar" between the fuze and the warhead digs in, lifting the base of the rocket upward while driving the fuze into the target and functioning it. Upon fuze functioning, an ejection charge is initiated, which ignites a pyrotechnic delay as it throws or bounds the warhead up and away. The pyrotechnic delay initiates the warhead 2 to 4 ft away from the point of impact.

Figure 7.12 Chinese type 69 bounding fragmentation rocket displayed with the shipping cap in place on the base and the unattached launch motor with fin assembly. (Author's photograph.)

General identification features associated with HE bounding rockets include:

- **Appearance and materials:**
  - The overall appearance is clumsy as if the rocket will not fly true.
  - Multi-piece body.
  - A motor is on the aft end with the warhead on the forward end.
  - Fin assembly on the aft end of the motor.
  - One or more venturis on the base of the motor.

- **Markings:** A green body with black markings is common. A brown band may be present on the motor. Other colors, stamped or stenciled markings, and symbols may also be present.
- **Common fuze configurations:** PD fuze for the ejection charge that ignites a very short pyrotechnic delay that initiates the bounding section of the warhead.
- **General safety precautions** for HE bounding rockets include:
  - HE, frag, movement; EMR, static, and ejection for an unfired motor.
  - Wait time (W/T) if the ejection charge deployed the warhead, but failed to function.
  - Safety precautions for the fuze if present.

1c. Rocket, HE thermobaric: Often mistakenly classified as "incendiary," these warheads are designed to explode and produce extremely high blast pressures in enclosed spaces. Employing liquid or jell-like main charges, thermobaric warheads are not very common. The obsolete U.S. M74 rocket, displayed in Figures 7.13 and 7.14, contains Tri-Ethyl Aluminum (TEA) housed in a thin aluminum body with a burster running down the center. The Russian RPO disposable rocket launcher has three different warheads and is defined as an "infantry rocket flame weapon." The only way to know which warhead is present in a launcher is a marking on the front cover plate. The RPO-A thermobaric warhead has two red stripes (Figure 7.15). The RPO-D smoke has one yellow

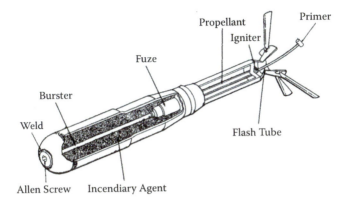

Figure 7.13 Line drawing depicting the internal configuration of the M74 TEA rocket. (From U.S. military TM.)

Figure 7.14 The actual color scheme of the M74. (Author's photograph.)

Figure 7.15 The special markings designating an RPO launcher as containing the thermobaric warhead. (Author's photograph.)

Figure 7.16 Russian RPO disposable rocket launcher, with the RPO-A thermobaric warhead and launch motor. (U.S. government.)

stripe and contains red phosphorus. The RPO-Z incendiary warhead has one red stripe and contains "pyrogel" that burns at 800°C to 1000°C.

General identification features associated with thermobaric rockets include:

- **Appearance and materials:**
  - A thin-skinned aluminum warhead.
  - Color codes that are inconsistent with common schemes.
  - A motor on the aft end with the warhead on the forward end.
  - Fin assembly on the aft end of the motor.
  - One or more venturis on the base of the motor.

Note: The launch motor may disconnect from the munition during flight.

- **Markings:** Color codes used for thermobaric rocket warheads are a departure from most standards, as seen on the anodized red body of the M74 or the unpainted body and gold ogive of the RPO (Figures 7.14 and 7.16). A brown band may be present on the motor. Other colors, stamped or stenciled markings, and symbols may be present.
- **Common fuze configurations:** BD fuze.
- **General safety precautions** for HE Bounding rockets include:
  - HE, frag, movement, fire, chemical; EMR, static, and ejection if motor is unfired.
  - The liquid fillers of thermobaric warheads usually pose a chemical threat.
  - Safety precautions for the fuze if present.

2. Rocket, HEAT: HEAT rocket warheads are configured in the same manner as HEAT projectiles with the addition of a motor. Surface-fired HEAT rocket warheads are usually made of lightweight aluminum materials (Figure 7.17), but older models were made of heavyweight steel. HEAT

Figure 7.17 The internal configuration of a surface-to-surface PG HEAT warhead; note the thin aluminum warhead design. (Author's photograph.)

rockets fired from aircraft also incorporate lightweight materials in warhead designs, but oftentimes include an anti-personnel fragmentation liner to increase peripheral lethality (Figure 7.18).

General identification features associated with HEAT rockets include:

- **Appearance and materials:**
  - Lightweight aluminum on newer models, heavy steel warheads on older models.
  - A break in the major diameter that may be forward of the bourrelet (a bourrelet is not always present on rocket warheads).
  - A motor on the aft end with the warhead on the forward end.
  - A hollow ogive crimped or screwed to the body that may have spanner holes.
  - Venturi(s) on the base of the motor.

- **Markings:** A green, black, or gray body with yellow or black markings is common. A brown band may be present on the motor. Other colors, stamped or stenciled markings, and symbols may be present.
- **Other:** Unlike HEAT projectiles, HEAT rockets do not commonly incorporate a standoff spike design as seen in Figures 4.6 and 4.36.
- **Common fuze configurations:** BD, PIBD, electric with piezoelectric (PE), and PIBD mechanical fuzing.

Figure 7.18 Russian S5KO air-to-surface HEAT rocket with an external coil-spring design fragmentation sleeve. (Author's photograph.)

- **General safety precautions** for HEAT projectiles include:
  - HE, frag, movement, jet, EMR, static, and ejection for an unfired motor.
  - Safety precautions for the fuze if present.

Note: Always assume a HEAT rocket has a PE fuze until proven otherwise, and include PE, EMR, and static for the fuze.

Note: Many rockets, such as the U.S. M72 LAAW, incorporate a PE fuze with a cocked striker (C/S) mechanical backup. The Russian PG series incorporates a pyrotechnic self-destruct (S/D) feature.

3. Rocket, dispenser: A hollow warhead with a payload that is sealed within. Depending on the design, the payload can be expelled from the nose or base of the warhead. With the rocket motor fixed to the base of the warhead, most rocket dispensers are configured for a forward deployment. As with projectiles, rocket dispensers containing HE or HEAT submunitions are treated as HE munitions. Examples of rocket-deployed submunitions are provided in Chapter 9 (Figures 9.9–9.11). After a rocket containing submunitions has functioned as designed, all that should remain is an empty rocket motor and an empty warhead body. Once ejected, the payload is categorized as a submunition since it has been separated from the rocket.

General identification features associated with rocket dispensers include (Figure 7.19):

- **Appearance and materials:**
  - A thin-skinned steel, aluminum, or fiberglass warhead.
  - A threaded or pressed nose cone that may be plastic.
  - Spanner holes or shear pins that may be present on the nose cone or near the warhead–motor junction.
  - Motor on the aft end with the warhead on the forward end.
  - Fin assembly on the aft end of the motor.
  - One or more venturis on the base of the motor.

- **Markings:** The payload will define the color codes and markings. A brown band may be present on the motor. Stamped or stenciled markings and symbols may be present.
- **Common fuze configurations:** MT, ET, and PTTF fuzing
- **General safety precautions** for rocket dispensers include:
  - Movement and ejection for the warhead; EMR, static, and ejection for an unfired motor.
  - Safety precautions for the fuze if present.

Figure 7.19 Yugoslavian 262mm, surface-to-surface, rocket dispenser warhead, which deploys the DPKB-1 HEDP submunition (Figure 9.9). (Courtesy of Dan Evers.)

Note: Depending on the payload, HE, frag, jet, WP, fire, and chemical may apply.

Note: The payload is explosively deployed and offers a substantial ejection hazard.

Note: All time fuzes require a W/T safety precaution.

Note: All MT fuzes require a C/S safety precaution.

4. Rocket, bursting smoke, WP: Smoke rockets are designed to produce smoke for screening, marking targets, or destroying material with fire. As seen in Figures 7.20(a) and 7.20(b), WP rocket warheads have white phosphorus sealed in the warhead by a burster adapter in a configuration similar to a type 1, WP projectile (Figures 4.51 and 4.52 in Chapter 4). Other designs include a modified HE or HEAT warhead with a base fuze such as Figure 7.21 and the RPO-Z warhead, which is similar in appearance to the RPO-A seen in Figure 7.16. When the fuze on a bursting smoke warhead functions, it initiates the burster, which detonates, breaking the warhead into pieces while dispersing the jelly-like WP.

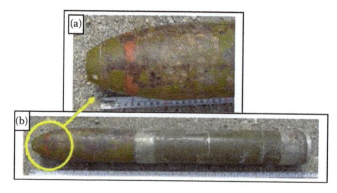

Figure 7.20 (a) A 107mm WP rocket with a nose fuze adapter booster. Note the identifiable features that include red markings and the top-down spanner holes on the adapter. (b) Identifiable features on 107mm WP rocket. (Author's photograph.)

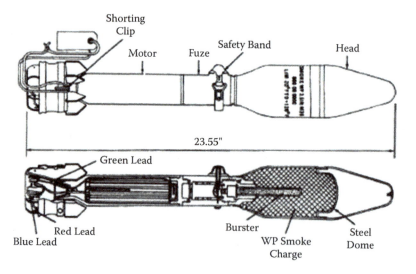

Figure 7.21 Line drawing of a U.S. 3.5 in. WP rocket with base adapter and fuze. (From U.S. military TM.)

General identification features associated with bursting smoke rockets include:

- **Appearance and materials:**
  - Other than the adapter booster, the warhead is of similar shape and size to that of HE or HEAT warheads.
  - Adapter booster may have wrench flats or spanner holes.
  - On non-U.S. ordnance, top-down or side spanner holes on the adapter may indicate whether the projectile is HE or WP (Figure 7.20).
  - A motor on the aft end with the warhead on the forward end.
  - Fin assembly on the aft end of the motor.
  - One or more venturis on the base of the motor.

- **Markings:** A lime green, olive drab, or gray body with yellow and red markings is common. A brown band may be present on the motor. Other colors, stamped or stenciled markings, and symbols may also be present.
- **Common fuze configurations:** PD or BD fuzing.
- **General safety precautions** for bursting smoke rockets include:
  - HE, frag, movement, WP, fire and chemical; EMR, static, and ejection for an unfired motor.
  - Safety precautions for the fuze if present.

Note: If burning, WP smoke is toxic.

5. Rocket, illumination: A dispenser type warhead design that ejects a parachute and pyrotechnic candle to illuminate an area at night. Illumination warheads contain at least two fuzes (Figure 7.22). One fuze is to eject the payload while the second fuze ignites the candle; some designs include additional mechanisms to rapidly deploy the parachute.

General identification features associated with illumination rockets include:

- **Appearance and materials:** All the construction features associated with dispenser type rockets.
- **Markings:** A white, green, or gray body with black, red, or brown markings is common. A brown band may be present on the motor. In Figure 4.19 (Chapter 4), the umbrella-like symbol is a common international marking used to distinguish illumination ordnance. Other colors, stamped or stenciled markings, and symbols may be present.
- **Common fuze configurations:** MT, PTTF, and ET fuzing.
- **General safety precautions** for illumination rocket include:
  - Movement, fire; EMR, static, ejection, as well as ejection for an unfired motor.
  - Chemical if the candle is burning.

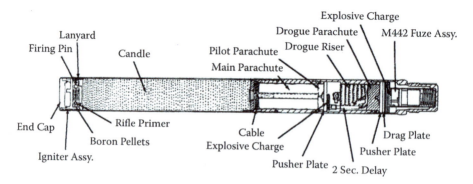

Figure 7.22 Line drawing of a U.S. 2.75 in. M257 multifuzed illumination rocket warhead. (From U.S. military TM.)

- Safety precautions for the fuze if present.

Note: The payload is explosively deployed and offers a substantial ejection hazard.

Note: Many designs include internal fuzing.

6a. Rocket, practice, with and without spotting charges: Are designed to be fired with the same ballistic characteristics as the live rockets they mimic, without the same destructive effects. This category includes subcaliber practice rockets fired through modified weapon systems. Practice rocket warheads may be solid metal and contain inert filler, such as powder dye, or a spotting charge propelled by a substantial explosive charge initiated by a live fuze. All designs include a live rocket motor (Figure 7.23).

General identification features associated with practice rockets include:

- **Appearance and materials:**
  - Construction features associated with the rocket they are designed to imitate.

Figure 7.23 U.S. 3.5 in. practice rocket with a live motor and ID features consistent with a HEAT warhead. (Author's photograph.)

- Designed for cost savings and training in restrictive settings, subcaliber practice rockets share no appearance similarities with the rocket they mimic.

- **Markings:** A blue or black body with white markings is common. A brown band may be present on the motor. Other colors, stamped or stenciled markings, and symbols may be present.
- **Common fuze configurations:** Depend on rocket type.
- **General safety precautions** for practice rockets include:
  - Movement; EMR, static, and ejection for an unfired motor.
  - HE, frag, and ejection when a spot charge is present.
  - Observing all applicable safety precautions for the live projectile until positive identification is made.
  - Safety precautions for the fuze if present.

Note: Practice means "practice"—not "inert."

6b. Rocket, drill and dummy are not designed to be fired and contain no energetic materials whatsoever. Drill and dummy rockets are used for weapon system loading and unloading drills or display.

General identification features associated with drill and dummy rockets include:

- **Appearance and materials:** Construction features associated with the rocket they are designed to imitate.
- **Markings:** A gold, black, or blue body with white markings is common. A brown band may be present on the motor. Other colors, stamped or stenciled markings, and symbols may also be present.
- **Common fuze configurations:** should not have a fuze.
- **General safety precautions** for practice rockets include:
  - Movement, EMR, static, and ejection until positive identification is made.

Closing

All ordnance is inherently dangerous, but unfired rockets possess an ability to launch and require additional threat considerations. Until proven otherwise, always consider a rocket to be in a hazardous condition.

Ordnance Category—
Guided Missiles

8

> Victory smiles upon those who anticipate changes in the character of war, not upon those who wait to adapt themselves after they occur.
>
> **General Guilo Douhet (Italian), 1921**

Introduction

General Douhet's comments almost 100 years ago can be applied to numerous advancements in military ordnance. The rapid technological advancements of sophisticated Guided Missile (GM) systems over the last few decades make this task more difficult from a number of perspectives, especially recognizing the hazards associated with an unknown missile found outside military control.

The history of GM development is an offshoot of rockets as simply defined; a GM is just a rocket with an ability to change its path during flight. All the information concerning propellants and rocket motors covered in previous chapters also applies to missiles, except that the motors are referred to as "missile motors" when affixed to a missile.

For this chapter, the defining factors that categorize a munition as a "Guided Missile" are that the body being projected:

1. Is propelled by a missile motor or motors. This definition is easier to apply if the motor remains attached to the warhead during flight, but this is not always true.
2. The munition is internally or externally guided and capable of altering its trajectory while in flight.

Missile Types

Many missile designs can be deployed from more than one platform. The first defining characteristic of a missile is its "Type," a classification derived from the launch point and the nature of the intended target. Missile types include:

1. **Surface-to-Surface** missiles are fired from the surface (ground or water) against targets such as people, vehicles, or ships that are also on the surface.

195

2. **Surface-to-Air** missiles are fired from the surface (ground or water) against airborne targets, including fixed-wing aircraft, helicopters, unmanned drones, rockets, or other missiles.
3. **Air-to-Surface** missiles are fired from manned or unmanned airborne aircraft against ground- or water-based targets such as people, vehicles, or ships that are on the surface.
4. **Air-to-Air** missiles are fired from manned or unmanned airborne aircraft against other airborne targets, including fixed-wing aircraft, helicopters, unmanned drones, rockets, or other missiles.

Key Identification Features

If the missile type can be ascertained, it will help in the identification process. A missile is a long, cylindrically shaped munition with a warhead and one or more motors containing one or more venturis and steerable fins that allow the munition to change direction in flight. Missiles are usually manufactured to higher standards than those of other ordnance categories and tend to incorporate light, yet strong materials such as aluminum, fiberglass, and plastic. As such, the quality of the materials used and the craftsmanship applied may be evident.

Note: There are exceptions to every rule and some missiles are guided by means other than steerable fins. For example, after launch, the US M47 Dragon GM is propelled and steered by 66 small, side-mounted motors, which can be seen in the bottom of Figure 3.11.

Missile Sections

Guidance section: Early versions of missiles were externally guided, but technological advancements allowed guidance systems to be placed on the missile itself. Guidance options include TV, laser, radar, infrared, and wire. Depending on the design, the guidance section can be located in the front, rear, or toward the middle of a missile. Figures 8.1 and 8.2 show examples of front- and rear-configured guidance sections.

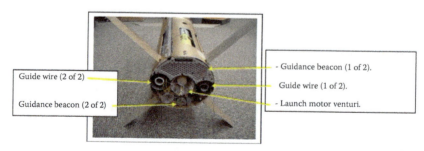

Figure 8.1 U.S. TOW-2A HEAT guided missile (GM). (Author's photograph.)

Figure 8.2 Russian SA-16 surface-to-air missile in the launch tube. The seeker head portion of the guidance section is damaged. (Author's photograph.)

Examples of guidance that involve a person include:

- Wire guided: A person controls the missile by providing flight path changes through wire that unspools throughout the missile's flight (Figure 8.1).
- Target designator: A person tracks a target and guides the missile to it. Any means of "showing" a missile where the target is can serve as a designator, including radar and lasers. The designator provides the information required for the GM to alter its direction to intersect with the designator on the intended target.

Examples of guidance that removes a person from the equation include more sophisticated systems—sometimes simplistically termed "fire and forget"—that remove the person as soon as the missile is aimed or programmed and fired. Some systems follow a predetermined route, while others are capable of differentiating between targets and, in many cases, select which target to go after. Examples include:

- Terrain mapping: Key terrain features seen by the missile during flight are compared to a programmed flight path. Course corrections are made to keep the missile on track until it reaches the target.
- Anti-aircraft guidance systems: Many of these systems are aimed at a target to lock on so as to "see" it, develop an intercepting trajectory, and tell the person controlling the system when this process is complete; the operator then launches the missile (Figure 8.2).

Note: If for some reason there is more than one potential target for an anti-aircraft missile to choose from, or if the missile loses, then reacquires a target after launch, the missile may decide to go after a target other than the one selected by the operator. When damaged, the complex electronics associated with missile systems capable of differentiating between different targets offer numerous hazards.

Control section: This section includes electrical and mechanical components that take information from the guidance section and convert it into adjustments for movable portions of the fins to alter the missile's flight path (Figures 8.3 and 8.4).

Fin assemblies: Missiles usually have more than one set of fins, with one fixed set and a second set that is movable or has movable sections. Fins can also pose an ejection hazard as some open with tremendous force. Others complete circuits upon opening and play a part in the fuze-arming process. If present, fins offer a differentiating identification feature between a missile and a rocket. Many missile systems will have multiple versions of the same missile and a specific fin design may be the only external means of telling them apart.

Motor section: Some large missiles are propelled by a jet engine, but these designs are usually very large and seldom encountered outside military control. The missile motors discussed in this chapter include designs that are similar to rocket motors, which consist of a body, igniter, propellant, and one or more venturis. Many missile designs contain more than one motor, in

Figure 8.3 Steerable fins mounted on the control section of a U.S. high-speed anti-radiation missile (HARM). (Author's photograph.)

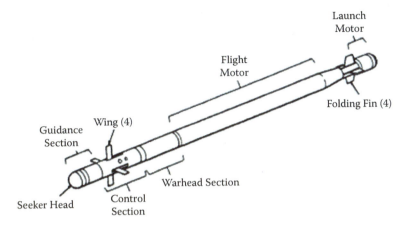

Figure 8.4 Line drawing of a U.S. Stinger, surface-to-air missile. Note: After initial deployment, the launch motor falls away allowing unobstructed functioning of the flight motor. (From U.S. military TM.)

which case a common configuration is a "launch" motor to initially deploy the missile and a "flight" motor to allow the munition to travel at a sustained speed for a specific amount of time.

Depending on design, the base (aft end) of the missile may have a motor venturi (Figure 8.1). Some motor configurations, such as the US Stinger or Russian SA-16 have a launch motor that detaches from the missile after initial deployment as it will obstruct the flight motor venturi (Figures 8.4 and 8.5). Other designs, such as the TOW (Tube-launched, Optically-tracked, Wireguided) missiles have a flight motor with side-venting venturis (Figure 8.6).

Figure 8.5 Launch motors from a Russian SA-16 surface-to-air missile. Left: unfired; right: fired. The motors on an SA-16 are set up in the same configuration as the Stinger missile (Figure 8.4).

Figure 8.6 U.S. TOW, HEAT missile. Note: The launch motor can be seen in Figure 8.1. The flight motor is the black section circled in yellow, and the black oval just aft (or left) of the motor is one of two venturis for the flight motor. (Author's photograph.)

When inspecting a missile, avoid areas in front of venturis. To determine if a motor has been fired, use a mirror to inspect the venturi(s) and establish if a closure disk or plug is missing. Figure 8.5 (left) shows closure disks in place and (right) closure disks blown. Count the number of venturis, measure their diameter, and note the venturi types (i.e., whether they are straight, slanted, canted or fluted) as this information will greatly assist in the identification process.

Warhead section: The warhead defines the group to which a missile belongs. Due to the way missiles are constructed, identification characteristics consistent with projectiles or rockets may or may not be found on missiles.

Warning: If a missle warhead is found seperated from an intact missile, apply all of the safety precautions associated with the missile that it is suspected to be until positive identification is made.

Fuze section: Some missile fuzing configurations are different from fuzing used in other ordnance for a number of reasons, one of which is that the target may be in constant motion. For example, an anti-aircraft missile traveling at over Mach 2 closes on an aircraft traveling toward it at Mach 1. If the target aircraft makes a flight path change that the incoming missile cannot counter and will be unable to impact the target, the missile will rely on VT fuzing to self-destruct the warhead at a location that will ensure the target is damaged or destroyed. At Mach 3 the window of opportunity to accomplish this is very small, requiring extremely accurate fuze functioning.

For this reason many missiles incorporate complex fuzing systems and Safe and Arming (S&A) devices to guard against malfunction and increase accuracy tolerances. As with all ordnance categories and groups, missile fuzing is usually designed to require three or more actions to arm and these actions coincide with the way a missile is deployed—for example, prior to or upon firing, a missile may require:

1. The operator remove pins, clips, or electrical shunts and turn on all power to the launch platform. Then the missile is locked onto its target and fired.
2. Upon firing, launch motor initiation results in (1) **setback** from the acceleration, followed by (2) **deceleration** as the missile slows during flight, and (3) **setback** when the flight motor fires, releasing locks or clockwork mechanisms, initiating pyrotechnic delays, and other actions to complete the arming sequence.

As with all munitions designed to be used for anti-aircraft functions, the fuze contains a Self-Destruct (S/D) feature. Anti-aircraft munitions are usually employed over friendly forces and if the missile fails to intercept its target, the S/D ensures that an intact, functional missile does not fall to the ground and detonate. By self-destructing, the missile will fall to the ground in small pieces that are less likely to cause death or injury. Additionally, most missile fuzes are internal to the munition and offer few clues to their existence. As such, the possibility of hidden or unseen fuzing must always be considered.

Additional hazards: The complex design of missiles oftentimes includes substantial hazards, including high-pressure gas bottles, generators for hydraulic power, high-voltage thermal batteries, and capacitors. These are in addition to toxic compounds such as mercury thallium and radiological sources used in many guidance sections. Until these hazards are identified and deemed safe, or their presence is ruled out, consider applying a chemical safety precaution (Figure 8.7).

Figure 8.7 A helium bottle in a U.S. TOW missile. (Author's photograph.)

The Seven-Step Practical Process Applied to Missiles

Examples of different designs, color codes, markings, and construction features are provided throughout this section.

Step 1: Approach and initial interrogation. Attempt to identify a munition at a distance with the use of binoculars. If an approach is made, avoid all venturis and fuze-sensing elements. Armed and active, or damaged sensing elements may "see" a person approaching, consider the person a valid target, and function as designed. At a minimum, measurements must be taken of the major diameter and the overall missile's length. If possible, separate diameter and length measurements of each section should be taken. Look for stamped data on the munition. All findings, including measurements, color codes, markings, key identifying features, and any possible damage, should be documented and the munition photographed.

In addition to the overall configuration, there are four features that will greatly assist in answering steps 2, 3, 5, and 7:

1. The diameter of the warhead and motor.
2. The overall length.
3. The locations and number of venturis (fired or unfired).
4. Which fins are movable or have movable surfaces.

Step 2: Determine fuze type and condition. If a missile has been deployed, the fuze is considered to be armed (step 5). If a fuze is damaged, pins have been removed, or any alterations have been made to the munition, it is considered armed. If visible, fuze measurements are taken separately from the munition.

Step 3: Determine ordnance category. This category covers missiles, which are fired by a propellant-filled motor as the primary means of deployment. Missiles incorporating jet engines are very large and seldom recovered outside military control. A brown-colored band may indicate the presence of a missile motor.

Step 4: Determine ordnance group. Identifying characteristics associated with each missile group will be covered throughout this chapter, many of which are consistent with the grouping characteristics of other categories.

Step 5: Determine if the munition was deployed. Inspect the forward end of the missile for impact-related damage. Inspect the venturi(s) for the presence of a closure disk or plug; if one is present, the motor has not been fired. If the motor has been fired, the closure disk(s) or plug will be missing.

Step 6: Determine safety precautions that apply to the munition. The safety precautions for missile groups are covered in this chapter. Chapter 2 addresses safety precautions associated with various fuzes. Due to the complex electronics inherent in missiles, they are highly susceptible to electromagnetic radiation (EMR) and static when damaged. Also apply EMR, static, and ejection to all unfired missile motors.

Note: Adhere to all safety precautions that apply to the missile and fuzing.

Step 7: Identify the munition. Apply the totality of all construction characteristics and other identifying features to determine the group to which a missile belongs and, if feasible, positively identify the munition and all possible fuzing configurations.

Groups

Per the definition provided in the Introduction section of this chapter, the missile category encompasses thousands of different ordnance items. In order to provide a coherent flow, the missile category is divided into the following primary and supplemental groups:

1. HE/frag.
2. HEAT.
3. Dispenser.
4. Practice.
   a. With active components.
   b. Drill and dummy.

1. Missile, HE/frag: Missile designs within this group are substantially different from the HE/frag warheads consistent in other categories. These missiles contain high-explosive warheads that explode to produce destructive effects through blast pressures and/or fragmentation, but the way in which they achieve these effects is oftentimes unique.

For example, anti-aircraft missiles have HE warheads ranging in size from a soda can with 12 oz (355 mL) of explosives found in a MAN Portable Air Defense System (MANPADS) such as the Stinger (Figure 8.4) to missiles with barrel-size warheads containing hundreds of pounds of HE such as the truck-launched SA-2 (Figure 8.9). MANPADS are designed to attack low-flying aircraft, while large truck-launched Surface-to-Air Missiles (SAMs) are designed to target high-flying aircraft; however, they both share a common characteristic. Anti-aircraft munitions are usually designed to function when they impact a target, but if the missile is going to miss, the next best choice is to self-destruct as closely to the target aircraft as possible. As the latter option is likely, missile warhead designs often involve multiple layers of unique fragmentation sleeves or liners designed to saturate the air near the target with razor sharp fragments. Figure 8.8 offers an example of these warheads showing a Continuous

Figure 8.8 CROW warhead from an air-to-air missile. (Author's photograph.)

Figure 8.9 The author inspecting a damaged SA-2 missile in Kuwait during the First Gulf War.

ROd Warhead (CROW) from a U.S. Sidewinder air-to-air missile. Upon detonation, multiple layers of approximately 1 ft long (305mm) diamond-shaped rods are explosively projected outward like spinning buzz saws.

One example of an air-to-surface or surface-to-surface missile warhead that does not rely on the effects of fragmentation is an anti-ship missile. Designed to function after penetrating the side or deck of a ship, many of these warheads include anti-deflection features normally associated with HEAT projectiles (see Figure 4.36 in Chapter 4). If the missile impacts a ship at an angle and is unable to penetrate, this feature will ensure the warhead is not deflected and functions as closely to the target as possible.

General identification features associated with HE/frag missiles include:

- **Appearance and materials:**
  - Missile and warhead are usually multi-piece construction.
  - Fuze is usually not observable and may not have any markings.
  - There are more than one set of fins, one of which is movable or has movable surfaces.
  - A motor is on the aft end with a possible second motor midbody.
  - One or more venturis are on the base of the motor or side of the missile body.
  - The nose is glass or plastic.
  - Guidance components such as wire spools or glass on the base are present.

- **Markings:** Green, white, gray, and black bodies with yellow, white, or black markings are common. A brown band may be present on the motor(s). Other colors, stamped or stenciled markings, and symbols may also be present.
- **Common fuze configurations:** Electrical impact and/or VT fuzing with self-destruct (S/D). Many anti-aircraft VT configurations include side-mounted proximity sensors versus the plastic nose arrangement consistent with many other ordnance categories.
- **General safety precautions** for HE/frag rockets include:
  - HE, frag, movement.
  - EMR, static and ejection for an unfired motor and ejection for fins if applicable.
  - Chemical for additional hazards and some propellant fuels if applicable.
  - Safety precautions for the fuze if present.

2. **Missile HEAT:** A very common warhead on surface-to-surface and air-to-surface missiles, HEAT warheads are configured the same as HEAT munitions of other categories. Figure 8.10 depicts a US TOW missile and

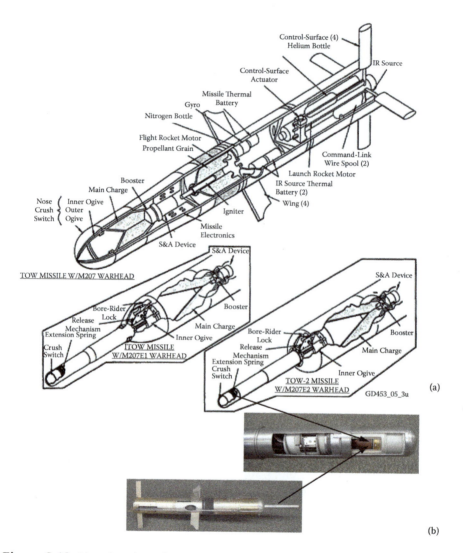

Figure 8.10 Line drawing of a U.S. TOW, HEAT missile series. (From U.S. military TM.); (a) Probe of a TOW-2A, with explosive charge. (Author's photograph.); (b) TOW-2A missile with fins deployed. (Author's photograph.)

three generations of warhead configurations, which offers an excellent example of how HEAT munitions have developed in order to circumvent countermeasures designed to defeat them. In Figure 8.10, the top-left "TOW" warhead incorporated a simple Point-Initiating Base-Detonating (PIBD) crush switch. Impact with the target crushed the thin outer ogive until it made contact with the inner ogive liner to complete a circuit and initiate the Base-Detonating (BD) element (Figure 3.11). The second generation, "Improved TOW" or "ITOW" (Figure 8.10, center) included a standoff probe that deployed during flight and increased the shaped charge's standoff from the target, which improved its penetration performance. But both of these warhead designs could be defeated by reactive armor, which disrupts jet formation as the warhead is detonating, resulting in the development of the "TOW-2A" (Figure 8.10, bottom; 8.10a; and 8.10b). Upon impact with a target equipped with reactive armor, the small shaped charge in the probe of the TOW-2A detonates, sympathetically detonating the reactive armor while initiating the BD element of the primary warhead charge (note wires at far left of Figure 8.10a). After the disruptive shock-front from the reactive armor passes, the main shaped charge functions, penetrating the target.

General identification features associated with HEAT missiles include:

- **Appearance and materials:**
  - Missile and warhead are usually multi-piece construction.
  - Fuze is usually not observable and may not have any markings.
  - There are more than one set of fins, one of which is movable or has movable surfaces.
  - A motor is on the aft end with a possible second motor midbody.
  - One or more venturis are on the base of the motor or side of the missile body.
  - There are guidance components such as wire spools or glass on base.
  - There is a break in the major diameter, behind the ogive.
  - A hollow ogive is riveted or screwed to the body.

- **Markings:** Green and black bodies with yellow or black markings are common. A brown band may be present on the motor(s). Other colors, stamped or stenciled markings, and symbols may also be present.
- **Common fuze configurations:** Electrical PIBD or BD with a S/D fuze:
- **General safety precautions** for HEAT missiles include:
  - HE, frag, movement, jet.
  - EMR, static, and ejection for an unfired motor and ejection for fins if applicable.
  - Chemical for additional hazards and some propellant fuels if applicable.
  - Safety precautions for the fuze if present.

3. Missile, Dispenser: As with other categories, a dispenser is a hollow warhead with a payload that is ejected over the intended target. Common payloads include HE and HEAT submunitions as well as dispensed land-mines. As with projectiles and rockets, a missile dispenser containing HE munitions is treated as an HE missile. Examples of missile-deployed submunitions are covered in a later chapter; however, once ejected; the payload is categorized as a submunition because it has been separated from the missile.

General identification features associated with missile dispensers include:

- **Appearance and materials:**
  - Missile and warhead are usually multi-piece construction.
  - Fuze is usually not observable and may not have any markings.
  - There are more than one set of fins, one of which is movable or has movable surfaces.
  - A motor is on the aft end with a possible second motor mid-body.
  - One or more venturis are on the base of the motor or side of the missile body.
- **Markings:** The payload will define the color codes and markings. A brown band may be present on the motor(s). Other colors, stamped or stenciled markings, and symbols may also be present.
- **Other:** Missiles from this group tend to be larger and launched from a vehicle or ship.
- **Common fuze configurations:** ET fuzing, which may include an S/D feature.
- **General safety precautions** for dispensers include:
  - Movement.
  - Depending on payload and warhead design, HE, frag, jet, and ejection may apply.
  - EMR, static, and ejection for an unfired motor and ejection for fins if applicable.
  - Chemical for additional hazards and some propellant fuels if applicable.
  - Safety precautions for the fuze if present.

4a. Missile, practice, with active components: Few practice missiles are designed to be fired due to the expense. In order to address training needs, many practice missiles are fitted with active guidance systems and tracking software allowing a gunner to simulate firing and have the accuracy of the "shot" tracked by computer software. Even though a live warhead and motor are not present, many of these systems still possess substantial hazards.

General identification features associated with practice missiles containing active components include:

- **Appearance and materials:**
  - The construction features associated with the missile they are designed to imitate.
  - Electrical connections on the body.
  - Fully functioning guidance section.

- **Markings:** A blue or yellow body with white or black markings is common. There are no yellow or brown bands. Other colors, stamped or stenciled markings, and symbols may also be present.
- **Common fuze configurations:** None.
- **General safety precautions** for practice missiles include:
  - Movement.
  - Observing all applicable safety precautions for the live missile until positive identification is made.
  - Chemical for additional hazards and some propellant fuels if applicable.
  - Practice means "practice"—not "inert".

4b. Missile, drill and dummy: Designed to be used for weapon system loading, unloading drills, or display and should not contain active components or energetic materials.

General identification features associated with practice missiles include:

- **Appearance and materials:** The construction features are associated with missile they are designed to imitate.
- **Markings:** Gold, black, and blue bodies with white markings are common. Other colors, stamped or stenciled markings, and symbols may also be present.
- **Common fuze configurations:** None.
- **General safety precautions** for practice missiles include:
  - Movement.
  - Observing all applicable safety precautions for the live missile until positive identification is made.

Closing

All ordnance is inherently dangerous, but unfired missiles possess an ability to launch and require additional threat considerations. The additional threats associated with guidance section, such as mercury thallium also warrant additional consideration

Ordnance Category—Submunitions 9

Known U.S. cluster bombs dropped during Operation Desert Storm amounted to 47,167 units containing 13,167,544 bomblets. It has been estimated that 30,000 tons of unexploded ordnance [UXO] was scattered across Kuwait when the Gulf War ended. By February 1992 [one year after the war ended] more than 1,400 Kuwaitis had been killed in incidents involving UXO and landmines. Among the most dangerous items were cluster bomblets.

> Rae McGrath, Cluster Bombs, Military Effectiveness and Impact on
> Civilians of Cluster Munitions
> (Landmine Action, 2000)

Introduction

Submunitions are some of the most contemptible munitions used on the battlefield. If found outside military control, submunitions constitute a multitude of observable dangers as well as many concealed threats as they are often designed to kill or disrupt those personnel trained to clear them. The statistics associated with submunition use in the 1991 Gulf War provided by Mr. McGrath represent dispensers or "units" dropped by U.S. aviation assets. Not included in these numbers are the projectile, rocket, and guided missile dispensers fired by U.S. forces or the dispensers deployed by all of the other countries involved in the fighting. Mr. McGrath is absolutely correct when stating that submunitions—also known as cluster bomblets—were among the most dangerous ordnance items endangering Kuwaiti civilians. Submunitions are generally small with no standardized shape or configuration; as such, they are easily overlooked, concealed, or secreted off as war trophies.

Basically defined, a submunition is an ordnance item deployed from a projectile, rocket, guided missile, or aerial dispenser. They are usually intended to saturate large areas with very simplistic fuzing designs or crafted to destroy specific targets with highly sophisticated tracking and fuzing systems. Both of them have very high "failure to function" or "dud" rates that often exceed 20% of the payload.

For this chapter, the defining factors that categorize a munition as a "Submunition" is that the munition:

1. Is a subcomponent of a munition classified under the Group "dispenser."
2. Is deployed from its dispenser as an independent, fully functional ordnance item.

211

There are exceptions to this definition and variations that may confuse accurate identification for a number of reasons. The vernacular used to describe or define submunitions is generated by the category of the dispenser from which they are deployed. As submunitions can be deployed from a number of ordnance categories, numerous naming conventions may be applied. For example, depending on the submunition being researched, it may be classified as bomblet, submunition, cluster munition, grenade, or dispensed landmine, as well as other terms. In order to avoid confusion, the term submunitions is used throughout this text. "Grenades" are defined as the hand, rifle, and projected grenades covered in Chapter 5, and landmines will be covered in Chapter 10. There is some gray area concerning the defining differences between submunitions and dispensed landmines. In an attempt to bring some clarity to a rather complicated classification, an example of a "family" of dispensed mines is covered in the landmines chapter.

Key Identification Features

The identifying characteristics of a munition are generated by the physics and engineering associated with the category and group to which it belongs, and this is sometimes true with submunitions. However, as a payload or sub-component of a larger munition, submunitions lack the characteristics consistent with being "fired" or "shot" in a conventional manner, and they are often devoid of any appropriate color codes or markings.

Submunitions come in many shapes and sizes and can be deployed in a number of different ways. If present, all of the definitions associated with identifiable ordnance characteristics, such as ogive, bourrelet, warhead, body, motor, fin assembly, etc., apply to submunitions. Unlike projectiles, rockets, and guided missiles, submunitions may not have the identifying components consistent with these other ordnance categories, thus hindering identification.

When determining the category of an unknown munition, the question to be answered is, "How did it get here?" If the answer to this question cannot be ascertained, an initial hypothesis that the item is a submunition would be a safe course of action until additional information becomes available.

The Seven-Step Practical Process Applied to Submunitions

Examples of different designs, features, color codes, markings, and construction features are provided throughout this chapter.

Step 1: Approach and initial interrogation. Attempt to identify a munition at a distance with the use of binoculars. If an approach is made, avoid venturis and fuze-sensing elements. Armed and active or damaged sensing elements may "see" a person approaching, consider the person a valid target, and function as designed.

At a minimum, measurements must be taken of the major diameter and the overall length of the munition. If different sections of the munition can be determined, measure the length and width of each. Look for stamped data on the munition. All findings, including measurements, color codes, markings, key identifying features, and any possible damage, are documented and the munition is photographed.

In addition to the overall configuration, there are four features that will greatly assist in answering steps 2, 3, 5, and 7:

1. The diameter.
2. The overall length.
3. A lack of specific components that assist in determining category (i.e., rotating bands, rocket motors, etc.).
4. A means of stabilization for the munition.

Step 2: Determine fuze type and condition. If a submunition has been deployed, the fuze is considered to be armed (step 5). If a fuze is damaged, pins have been removed, or any alterations have been made to the munition, it is considered armed. If visible, measurements for the fuze are taken separately from the munition.

Step 3: Determine ordnance category. This category covers submunitions deployed from a dispenser.

Note: Dispensed landmines are excluded from this definition.

Step 4: Determine ordnance group. Identifying characteristics associated with submunition groups will be covered throughout this chapter.

Step 5: Determine if the munition was deployed. Inspect the munition for impact-related damage, missing pins or clips, and deployed tripwires.

Step 6: Determine safety precautions that apply to the munition. The safety precautions for the submunition groups are covered in this chapter. Chapter 3 addresses the safety precautions associated with various fuzes.

Note: Some fuzing options and safety precautions will be covered that are specific to a munition.

Step 7: Identify the munition. Apply the totality of all construction characteristics and other identifying features to determine the group to which a submunition belongs and, if feasible, positively identify the munition and all possible fuzing configurations.

Groups

The submunition category encompasses thousands of different ordnance items. In order to provide a coherent flow, the submunition category is divided into the following primary and supplemental groups:

1. High Explosive (HE)
   a. HE/fragmentation (frag) and High-Explosive Incendiary (HEI)
   b. Bounding
   c. Fuel–Air Explosive (FAE)

2. High-Explosive Anti-Tank (HEAT) and Explosively Formed Projectile (EFP)
3. Incendiary
4. Practice

1. Submunition, HE and HEI: All of the configurations covered under this group are designed to explode and produce destructive effects through blast pressures and fragmentation. Explosive fillers range from less than an ounce to hundreds of pounds and may be solid, pliable, or in liquid form. Additionally, many HE submunitions contain pyrophoric metals such as aluminum, magnesium, or zirconium to produce an anti-material effect. The anti-material additives are so common that all HE submunitions are assumed to possess this hazard until proven otherwise. For this reason, HE and HEI submunitions are grouped together.

1a. Submunition, HE/frag, HEI: It is crucial to understand the initial purpose of these munitions to appreciate how they have evolved. The original submunition fell into the HE group, it was the German SD-2 "Butterfly" bomb (Figure 9.1), which was designed to destroy equipment, cause casualties, and, most importantly, disrupt operations. After deployment, the outer shell or "wings" opened to serve two purposes: (1) orient the munition, and (2) spin-arm the fuze. There are four fuzing possibilities for the SD-2 and each dispenser contains submunitions with all four fuze types:

1. Short time delay: To destroy equipment and cause casualties.
2. Impact: To destroy equipment and cause casualties.
3. Long time delay, self-destruct: To provide an area denial effect, hamper cleanup and recovery operations, and cause casualties.
4. Anti-disturbance (A/D): To hamper cleanup and recovery operations by specifically targeting those personnel tasked to recover UXOs.

Note: The mixed-batch fuzing approach developed for the SD-2 became a standard configuration for many submunition designs still used today (Figure 9.2–9.4).

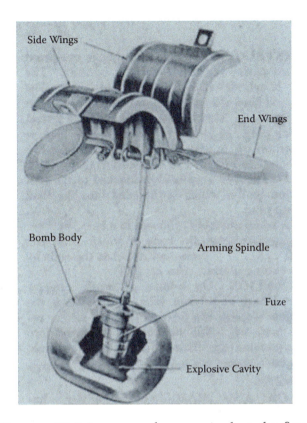

Figure 9.1 German SD-2 is commonly recognized as the first successfully fielded submunition incorporating a variety of fuzing options. The U.S. M83 is a "copycat" of this design and incorporates similar fuzing. A welded seam on the body of the U.S. version is an easy way to differentiate between the two. (From U.S. military TM.)

Figure 9.2 U.S. BLU 60 series submunitions, commonly referred to as the "golf ball," "baseball," and "softball" due to their similar sizes. After deployment, these munitions, as well as the Russian AO-2 (Figure 9.3), have external flanges that impart spin, which arms the fuze (Figure 9.4). (Author's photograph.)

Figure 9.3 Russian AO-2. (Author's photograph.)

General identification features associated with HE/frag, HEI submunitions include (Figures 9.1–9.4):

- **Appearance and materials:**
  - Heavyweight body appearance.
  - External serrations possibly visible.
  - Multi-piece body.
  - A means of arming.
  - A lack of orientation (the SD-2 design became an exception).
  - A lack of deployment characteristics (i.e., being "fired" or "shot").

Figure 9.4 With an internal fuze and few if any external markings, it is impossible to tell if the submunition contains an impact, short/long delay time, or an anti-disturbance fuze. Considering that the SUU-30 dispenser can carry as many as 3,000 "golf balls" and that an attack bomber can carry 16 or more dispensers, the number of dud submunitions present on a battlefield may be considerable. (Author's photograph.)

- **Markings:** The body may be unpainted or painted colors inconsistent with common schemes. Note any colors, stamped or stenciled markings, and symbols.
- **Common fuze configurations:** Internal fuzing that cannot be seen. Possibilities include all-way-acting impact, short time delay and long time delay Self-Destruct (S/D), A/D fuzing that may arm immediately after deployment, while in flight, or after landing on the ground.
- **General safety precautions** for HE/frag, HEI submunitions include:
  - HE, frag, movement, fire, Wait-Time (W/T), and Boobytrap (B/T) for the A/D feature.
- Electromagnetic Radiation (EMR), static, W/T, and B/T if electrical components are seen or suspected.

1b. Submunition, HE bounding: There are not many versions of this subgroup, and many of these designs are obsolete, but they are still recovered. A common design includes a means of orientation and pressure-plate. Immediately upon impact, the pressure plate is struck and initiates an ejection charge that ignites a pyrotechnic delay as it ejects or bounds the warhead upward, away from the submunition body. The pyrotechnic delay initiates the warhead 2 to 4 ft from the point of impact (Figures 9.5 and 9.6).

General identification features associated with HE bounding submunitions include:

- **Appearance and materials:**
  - Means of orientation (not found on all designs (Figure 9.7).
  - A multi-piece body

Figure 9.5 Two different designs of U.S. bounding fragmentation submunitions containing a warhead smaller than a golf ball. (Author's photograph.)

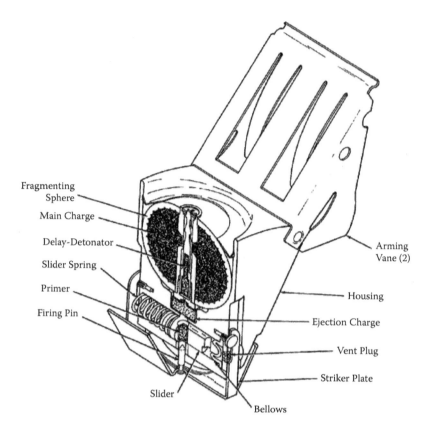

Fragmenting Sphere

Main Charge

Delay-Detonator

Slider Spring

Primer

Firing Pin

Arming Vane (2)

Housing

Ejection Charge

Vent Plug

Striker Plate

Slider

Bellows

Figure 9.6 Line drawing of a bounding submunition. (From U.S. military TM.)

Figure 9.7 The U.S. area denial anti-personnel mine (ADAM) is a good example of the "gray" area associated with submunitions and dispensed landmines. After impact with the ground, seven 20 ft (6.5 m) tripwires are deployed and a S/D timer initiated. If any of the tripwires or the body of the munition is disturbed, an ejection charge breaks open the resin body while initiating the fuze delay as it deploys the warhead, which detonates 4 to 5 ft (1.5 m) from the body. There is also a hand-thrown version, the M85 pursuit deterrent ADAM, which functions the same way. (Author's photograph.)

- **Markings:** May be unpainted or painted colors inconsistent with common schemes. Other colors, stamped or stenciled markings, and symbols may not be present.
- **Common fuze configurations:** PD, short pyrotechnic delay, A/D, and S/D.
- **General safety precautions** for HE bounding submunitions include:
  - HE, frag, movement, ejection.
  - W/T if the ejection charge deployed the warhead, but it failed to function.
  - EMR, static, W/T, and B/T if electrical components are seen or suspected.

1c. Submunition, Fuel–Air Explosive (FAE): Similar in construction to the thermobaric warheads found in other ordnance categories, FAE bomb deigns incorporate a time element that ensures a proper fuel–air balance, or stoichiometric ratio, for maximum explosive effect. Upon impact with the ground, a burster tube in the center of the munition functions, ripping open the thin-skinned body while aerosolizing the liquid fuel. As the fuel mixes with the air in the environment, "cloud detonators" arm and function as a stoichiometric ratio is reached (Figure 9.8). The result is a tremendous explosion and the generation of a strong shock-front designed to clear vegetation and allow helicopters to land in jungle and wooded terrain. FAE munitions are seldom used and the liquid fuel restricts storage times. They are seldom encountered outside military control.

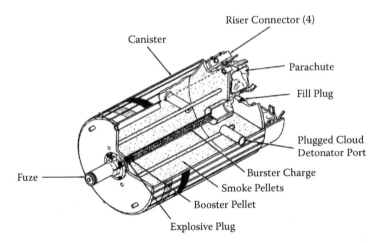

Figure 9.8 Line drawing of a U.S. BLU-73 FAE bomb. The CBU-72 dispenser contains three FAE bombs weighing approximately 100 lb (45 kg). Each munition explodes with the force of 250 lb (113 kg) of TNT. (From U.S. military TM.)

General identification features associated with FAE submunitions include:

- **Appearance and materials:**
  - A thin-skinned body.
  - May have a probe or standoff for the fuze.
  - A parachute of other means of orientation.
  - A filler plug of some kind on the body.

- **Markings:** The body may be unpainted or painted colors inconsistent with common schemes. Other any colors, stamped or stenciled markings, and symbols may be present.
- **Common fuze configurations:** Impact fuzing.
- **General safety precautions** for FAE submunitions include:
  - HE, frag, movement.
  - Fire and chemical as the fuels are volatile and some are toxic.

2. Submunition, High-Explosive Anti-Tank (HEAT) and Explosively Formed Projectile (EFP): Submunitions contain a shaped charge or EFP warheads to defeat armored and other hardened targets. Older HEAT submunitions were configured to maximize armor penetration, with little consideration of fragmentation production. Newer designs incorporate a "multipurpose" configuration to maximize all possible effects, including a shaped-charge or EFP for armor penetration, a fragmentation sleeve for anti-personnel, and a zirconium or other pyrophoric metal to produce an anti-material effect. These munitions often contain more than one fuze to ensure proper functioning against a variety of targets. These additional functioning characteristics lead to many submunitions being designated as multipurpose or High-Explosive Dual Purpose (HEDP), such as some projected grenades. For the purpose of this text, any submunition designed to defeat armor will be referred to as a "HEAT" submunition; when possible, additional warhead design characteristics will be mentioned. This group represents the most commonly deployed submunitions as well as the most commonly encountered outside military control. Figures 9.9–9.19 include examples from a number of designs that function very differently.

General identification features associated with HEAT submunitions include:

- **Appearance and materials:**
  - A means of orientation.
  - Heavy-body construction.
  - A break in the major diameter of the body (may not be observable).
  - Standoff spike.
  - A hollow ogive crimped or screwed to the body, which may have spanner holes.

Figure 9.9 Yugoslavian KB-1, HEDP submunition. (Author's photograph.)

Figure 9.10 U.S. M42, M46, or M77. HEDP configured submunition that can be artillery or rocket delivered. (Author's photograph.)

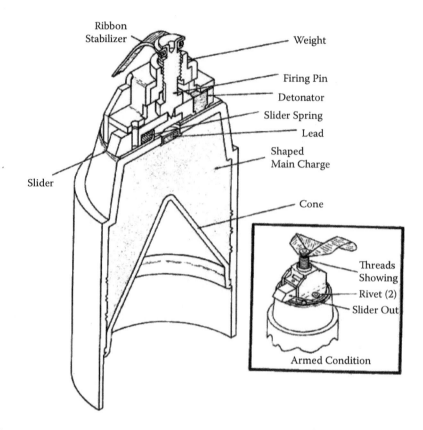

Figure 9.11 Line drawing of an M42. The KB-1 is similarly configured. These munitions are very simplistic and have very high dud rates. After arming, the submunition strikes the ground and impact inertia drives the firing pin into the detonator. (From U.S. military TM.)

- **Markings:** The body may be unpainted or painted colors inconsistent with common schemes. Other colors, stamped or stenciled markings, and symbols may be present.
- **Common fuze configurations:** Base Detonating (BD), Point-Initiating Base Detonating (PIBD), and long delay S/D fuzing.
- **General safety precautions** for HEAT submunitions include:
  - HE, frag, movement, jet.
  - PE (piezoelectric), EMR, static, W/T, and B/T if electrical components are seen or suspected.
  - Fire if a pyrophoric component is suspected.

Note: Always assume a HEAT munition has a PE fuze until proven otherwise and include PE, EMR, and static.

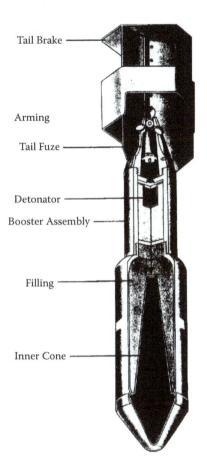

Tail Brake

Arming

Tail Fuze

Detonator

Booster Assembly

Filling

Inner Cone

Figure 9.12 Line drawing of a WWII era, Japanese Type-2 cluster bomb. (From U.S. military TM.)

Figure 9.13 Japanese Type-2 cluster bomb. Note the arming propeller with the fin and tail-boom assemblies meet. (Author's photograph.)

Figure 9.14 French GR-66 "Belouga" submunition, which may have an impact, time delay, self-destruct (S/D), or anti-disturbance (A/D) fuze. (Author's photograph.)

Figure 9.15 U.S. MK118 "Rockeye" HEAT with a PIBD (PE) fuze. Note the arming window on the fuze housing between the body and the fins. (Author's photograph.)

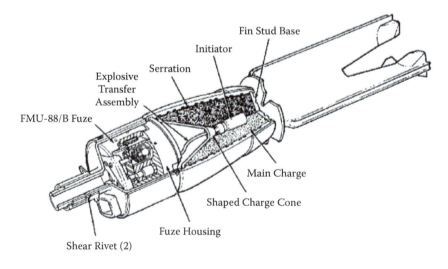

Figure 9.16 Line drawing of a U.S. BLU-77/B. (From U.S. military TM.)

Figure 9.17 The BLU-77/B is an older submunition defined as an Anti-Personnel Anti-Material (APAM), or a "multipurpose" munition. Upon impact with a hard target, the BLU-77 PIBD fuze will function and direct its shaped charge against the target. If a soft target such as sand or dirt is struck, an explosive charge in the fuze housing ejects the warhead into the air where the BD fuze functions, throwing fragmentation and zirconium, which is the yellow collar just forward of the fin assembly. (Author's photograph.)

Figure 9.18 U.S. BLU-108, an extremely sophisticated munition. After deploying from a dropped dispenser, a parachute controls the descent until a radar altimeter initiates the deployment sequence at a predetermined height. At that time the four submunitions are kicked out like skeet and use an infrared (IR) sensor to identify targets such as armored vehicles within their trajectory. When a target is identified, the warhead functions while the submunition is airborne (airburst) and an EFP is fired through the top of the target. If a target is not identified, the munition is designed to self-destruct or render itself safe. (Author's photograph.)

Figure 9.19 U.S. BLU-108.

3. Submunition, incendiary: Usually contains thermite or thermate mixtures that burn at approximately 4000°F, posing a serious fire hazard. These submunitions are designed to destroy equipment and material by melting or burning.

General identification features associated with incendiary submunitions include (Figures 9.20 and 9.21):

- **Appearance and materials:**
  - A means of orientation on some designs.
  - Light-body construction.
  - Multi-piece body.

- **Markings:** May be unpainted or painted colors inconsistent with common schemes. Note any colors, stamped or stenciled markings and symbols.
- **Common fuze configurations:** BD or PD fuzing that arms during flight; some designs are initiated when being deployed from the dispenser and fall to the ground already burning.
- **General safety precautions** for incendiary submunitions include:
  - Movement, fire.
  - Safety precautions for the fuze if present.

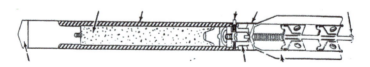

Figure 9.20 Line drawing of a U.S. M126. (From U.S. military TM.)

Figure 9.21 M126, a 4 lb (1.8 kg) incendiary bomblet. (Author's photograph.)

4. Submunition, practice is designed to be deployed with the same ballistic characteristics as the live submunition it mimics, but without the same destructive effects. Practice submunitions include designs that contain inert fillers or a spotting charge ejected by a fully functional fuze, while other designs are hollow versions of the live munition.

General identification features associated with practice submunitions include:

- **Appearance and materials:** Construction features associated with the submunition it is designed to mimic.
- **Markings:** The body may be unpainted or painted colors inconsistent with common schemes. Note any colors, stamped or stenciled markings, and symbols.
- **Common fuze configurations** may or may not be present depending on submunition type.
- **General safety precautions** for practice submunitions include:
    - Movement.
    - HE, frag, ejection when a spotting charge is present.
    - Observing all applicable safety precautions for the live submunition until positive identification is made.
    - Safety precautions for the fuze if present.

Note: Practice means "practice"—not "inert."

Closing

Submunitions do not always possess common ordnance identification features and many are easily mistaken for children's toys. When recognized, always consider a submunition to be in a hazardous condition until proven otherwise.

Ordnance Category—Landmines   10

Landmines do not recognize ceasefires or peace accords. They can go on killing decades after wars have ended. Anti-personnel mines have an average life span of 50–100 years. Over 400 million landmines have been laid since the beginning of World War II. Up to 80 million are still alive today in hundreds of thousands of minefields around the world.

The Landmine Site (hosted by Jeremy Gregg)

Introduction

Conventional landmine use as it is recognized today dates back to the American Civil War, when a design patented as "sub-terra boobytraps" by the Rains brothers was authorized for use by the Confederate secretary of war. With the Civil War raging and Confederate forces losing ground, the thought of deploying such an immoral or cowardly weapon restricted its employment to parapets and roads in order to slow or stop an assault. Though the nefarious reputation stuck, the restrictions did not and today landmines are some of the most commonly encountered ordnance items.

The unique aspect associated with landmines is that they are the only ordnance category specifically designed to be functioned by their intended victim. Landmines cause a number of tactical problems for opposing sides, but as pointed out on The Landmine Site, the biggest problems start when hostilities end as history has shown that just about all countries have a poor track record for retrieving the mines they laid.

Landmine designs vary and they can be hand placed, mechanically deployed from vehicles, or deployed from dispensers in the same manner as submunitions. With so many designs, it is impossible to cover every landmine, as such; this chapter focuses on shapes, construction features, materials, and functional designs that may help establish the group and safety precautions associated with an unknown landmine.

For this chapter, the defining factors that categorize a munition as a "Landmine" are that the munition:

1. Is designed to be initiated by the intended victim.
2. Is designed for land-based deployment.

There are ordnance configurations that can make this definition somewhat confusing and make accurate identification difficult. For example, after deployment and fuze arming, a 2,000-pound bomb dropped with an influence fuze such as the "Destructor" (Chapter 3) will function as a magnetic influence mine.

There is also some gray area concerning the defining differences between submunitions and dispensed landmines. In an attempt to bring some clarity to a rather complicated classification, an example of a family of scatterable mines (FASCAM) is provided throughout this chapter. For this purpose, the air-dropped "Gator," artillery delivered "Remote Anti-Armor Munition" (RAAM), and helicopter- or ground-placed "Volcano" systems will be used as all three deploy the same munitions.

Key Identification Features

Landmines are designed to be hidden or appear benign when deployed. As such, they come in many shapes and sizes and can be deployed in a number of different ways. Most landmine fuzes function when pressure is applied to a pressure plate on the top of the body. They are also assumed to be booby-trapped, which equates to additional hidden fuzing.

Landmine Sections

Body: Most landmine bodies are constructed of metal or plastic and may have a variety of different fittings, mounting brackets, and legs for deployment purposes. Other materials used for landmine bodies include composites, wood, or ceramics. Pressure plates are commonly affixed to the body; additional secondary and tertiary fuze wells are common on larger landmines.

Fuze: By design, a landmine is functioned by its victim. As such, the most common fuzing—both primary and secondary boobytrap type fuzing—involves pressure, pull, pressure release, tension release, or tilt generated by a person or vehicle. Other fuzing configurations include influence or electronic command fuzing. There are more sophisticated designs; for example, some landmine fuzes are capable of "sleeping" to save battery power. "Awakened" by seismic activity, a magnetic influence fuze scans the area for an appropriate target. If a target does not come within range during a preset time, the fuze may go back to sleep or self-neutralize and render itself inoperative. If the mine sits for so long that the battery voltage drops below a preset level, the fuze may initiate a Self-Destruct (S/D) feature.

Hand-placed landmines are usually direct armed by removing a pin or clip. As stated in Chapter 3, most large mines contain auxiliary fuze wells designed to be boobytrapped, which can lead to a tremendous amount of field improvisation. Always consider the possibility of hidden or unseen fuzing and treat any mine as if it is boobytrapped until proven otherwise.

The Seven-Step Practical Process Applied to Landmines

Examples of different designs, features, color codes, markings, and construction features are provided throughout this chapter.

Step 1: Approach and initial interrogation. **Do not approach a deployed landmine**. Attempt to identify a munition at a distance with the use of binoculars. If an approach is made, avoid all fuze-sensing elements as active, or damaged sensing elements may "see" a person approaching, consider the person a valid target, and function as designed.

At a minimum, measurements must be taken of the major diameter and height of the mine body. If numerous sections are present, measure each one and look for stamped data. All findings, including measurements, color codes, markings, key identifying features, and any possible damage, are to be documented and the munition photographed.

In addition to the overall configuration, there are four features that will greatly assist in answering steps 2, 3, 5, and 7:

1. The diameter.
2. The overall length.
3. Body construction and materials.
4. Obvious fuze configuration and material used (i.e., a rubber pressure plate on a plastic body).

Step 2: Determine fuze type and condition. If a landmine has been deployed, the fuze is considered to be armed (step 5). If a fuze is damaged, pins have been removed, or any alterations have been made to the munition, it is considered armed. If visible, measurements for the fuze are taken separately from the munition.

Step 3: Determine ordnance category. This category covers landmines designed to be hand-placed or deployed from a dispenser.

Step 4: Determine ordnance group. Identifying characteristics associated with landmine groups will be covered throughout this chapter.

Step 5: Determine if the munition was deployed. Landmines found outside controlled storage should be considered deployed. Inspect the munition for missing pins or clips and deployed tripwires.

Step 6: Determine safety precautions that apply to the munition. The safety precautions for landmine groups are covered in this chapter. Chapter 3 addresses the safety precautions associated with various fuzes.

Note: Some fuzing options and safety precautions specific to a munition will be covered.

Step 7: Identify the munition. Apply the totality of all construction characteristics and other identifying features to determine the group to which a landmine belongs and, if feasible, positively identify the munition and all possible fuzing configurations.

Groups

The landmine category encompasses thousands of different mines. In order to provide a coherent flow, the landmine category is divided into the following primary and supplemental groups:

1. Anti-personnel (APERS or AP).
   a. Blast.
   b. Fragmentation (frag).
   c. Bounding fragmentation.
   d. Directional fragmentation.

2. Anti-tank (AT).
   a. Blast.
   b. Armor-penetrating with shaped-charge or Explosively Formed Projectile (EFP).

3. Practice.

1. Landmine, AP or APERS: The landmine configurations covered under this group are designed to explode and injure or kill personnel. Explosive fillers in these landmines range from approximately 1 oz to 3 lb (28.35 to 1,361 grams). Fuze settings and types vary, but direct pressure of 5 to 35 lb (2.27 to 15.9 Kg) or a tripwire pull of 8 oz to 8 lb (.22 to 3.6 Kg) is common force requirements. APERS mines are designed to be deployed in high numbers to increase the possibility of success. They are also used in conjunction with AT mines as a deterrent to clearance operations.

1a. Landmine, AP, blast is a non-fragmentation-producing mine designed to target personnel with explosive blast effects. Most designs are deployed just below the surface of the ground. When functioned by a person stepping on the mine, the explosive force follows the easiest path of resistance and is focused upward away from the ground, maximizing the energetic potential against the target. AP blast mines are the most simplistic and inexpensive; they are also easily deployed and the most commonly encountered.

General identification features associated with AP, blast landmines include:

- **Appearance and materials:**
  - Deployed below the surface of the ground.

Figure 10.1 P4MK1, Pakistani APERS mine. The shipping cover (white plastic piece, left) is unscrewed and removed prior to deployment. (Author's photograph.)

- May be painted or constructed of colored plastic components.
- Multi-piece construction.
- A means of manual arming (removal of shipping cap may arm the mine).
- **Markings:** Tan, brown or green body with black markings is common. Other colors, stamped or stenciled markings and symbols may also be present (see "TS-50" indentation in Figure 10.2).
- **Common fuze configurations:** direct pressure. Pressure plate may be visible, but fuzing is usually internal. Self-destruct or anti-disturbance (A/D) fuzing is included on some designs.
- **General safety precautions** for AP, blast landmines include:
  - High explosive (HE), frag, movement, boobytrap (B/T)
  - Safety precautions for the fuze if present
  - Electromagnetic radiation (EMR), static, and wait time (W/T) if electrical components are seen or suspected

Figure 10.2 An Italian TS-50 in deployed configuration. The Italian VS-50 and Iranian YM-1 mines are (externally) almost identical. (Courtesy of Didzis Jurcins.)

1b. Landmine, AP, fragmentation: Designed to target personnel with blast and fragmentation effects. Most designs are deployed aboveground with tripwires to extend the mine's coverage area. Fuzing can be configured to function when tension is applied by a target hitting the tripwire, imparting "pull" on the fuze, or when the fuze is set up under tension and the tripwire is cut, allowing the fuze to function from the "tension release" (Figure 10.3). The AP version of the FASCAM mentioned in the introduction, the Gator, RAAM, and Volcano is the BLU-92/B, which is a fragmentation-producing dispensed landmine (Figure 10.4). With eight tripwires that can each deploy

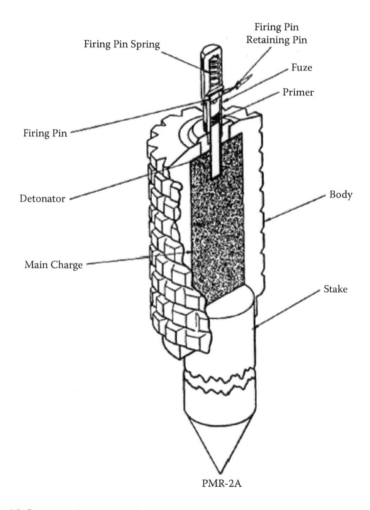

PMR-2A

Figure 10.3 Line drawing of a Yugoslavian PMR-2 is an example of an aboveground landmine with a thick, serrated body for fragmentation. It can be fitted with a tripwire used to initiate tension (pull) or tension release fuzing. (From U.S. military TM.)

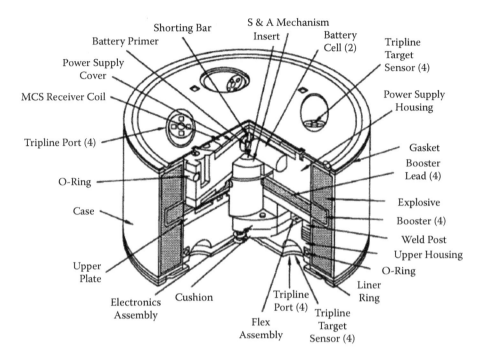

Figure 10.4 Line drawing of a U.S. BLU-92/B. APERS dispensed mine has four tripwires deploying from each side to ensure full area coverage. (From U.S. military TM.)

almost 20 ft (6 m) outward, these mines cover a large area. Containing almost 1 lb (0.4 kg) of high explosive, the fragmentation threat extends past the tripwire length.

General identification features associated with AP fragmentation landmines include:

- **Appearance and materials:**
  - Deployed above the surface of the ground.
  - Metal or plastic body.
  - Tripwires
  - Multi-piece construction.
  - A means of manual arming.
  - Multi-prong-like top on fuze (Figure 10.3).

- **Markings:** Tan, brown or green body with black markings is common for hand placed mines. Dispensed mines may or may not be painted. Other colors, stamped or stenciled markings and symbols may also be present.

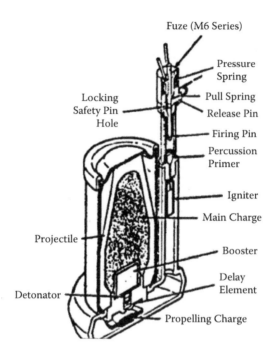

Figure 10.5 Line drawing of a U.S. M2 series bounding fragmentation landmine with an M6 series fuze. (From U.S. military TM.)

- **Common fuze configurations:** For a conventional design such as the PMR-2, tripwire(s) can be set for pull or tension release (Figure 10.3). For sophisticated designs such as the BLU-92/B, an internal A/D and low-voltage S/D may be employed in addition to the eight tripwires (Figure 10.4).

Note: In Figure 10.4, the mine contains a Safe and Arming (S&A) mechanism versus a conventional landmine fuze. Complex S&A devices are more commonly associated with missile fuzing.

- **General safety precautions** for AP fragmentation landmines include:
  - HE, frag, movement, B/T.
  - Safety precautions for the fuze if present.
  - EMR, static, and W/T if electrical components are seen or suspected.

Note: Do not cut tripwires.

   1c. Landmine, AP, bounding fragmentation: Deployed subsurface, these mines eject or bound the warhead approximately 2 to 4 ft (0.5 m) above the ground before it detonates. Commonly nicknamed the "Bouncing Betty,"

Figure 10.6 Cutaway of an M2 series. The warhead configured to bound out of the landmine body is a 60mm mortar set in a nose-down attitude with the fin assembly removed. (Author's photograph.)

this mine is not covered under fragmentation landmines due to its unique manner of functioning. The warhead and outer body are usually constructed of metal or plastic. Some designs incorporate a munition for its warhead, such as the 60mm mortar projectile used in the M2A3 (Figures 10.5–10.7), but explosive weights vary from a few ounces to over 7 lb (3.3 kg).

General identification features associated with bounding fragmentation landmines include:

- **Appearance and materials:**
  - Deployed above ground.
  - Metal body (may have plastic components).
  - Multi-piece construction.
  - A means of manual arming (removing pin from fuze where trip-wire attaches).

- **Markings:** Tan, brown or green body with black markings is common. Other colors, stamped or stenciled markings and symbols may also be present.

Figure 10.7 A deployed bounding fragmentation landmine. (Courtesy of Mark Ladd.)

- **Common fuze configurations:** Direct pressure and a tripwire set for pull and/or tension release. The prongs at the top of the fuze (Figure 10.5) will function the fuze when direct pressure is applied. Common designs include a pyrotechnic delay initiated by the ejection charge or a cable anchored to the body that will initiate a pull-friction fuze as the warhead bounds upward.
- **General safety precautions** for bounding fragmentation landmines include:
  - HE, frag, movement, ejection, and B/T.
  - W/T if the ejection charge deployed the warhead, but it failed to function.
  - Safety precautions for the fuze if present.
  - EMR and static if electrical components are seen or suspected.

Note: Do not cut tripwires.

1d. Landmine, AP, directional fragmentation: Are deployed above ground; These fragmentation-producing mines are designed to target personnel with blast and fragmentation effects. Commonly nicknamed "Claymores," these mines are not covered under fragmentation landmines due to their unique directional targeting ability. The outer body is usually constructed of metal or plastic. Designs vary, but all are configured with a layer of fragmentation, backed by explosives in a round or rectangular shape and a concave or convex front (Figures 10.8–10.10). Explosive weights vary from 1 to over 26 lb (0.4 to 12 kg).

Figure 10.8 Iranian copycat of the U.S. M18 Claymore mine. (Author's photograph.)

General identification features associated with directional fragmentation landmines include:

- **Appearance and materials:**
  - Deployed above ground.
  - Metal or plastic body.
  - Two-piece body; a front and back with a seam at the junction.
  - Legs or a spike to secure it in place.
  - A means of manual arming.

Figure 10.9 Russian MON100, side view. (Author's photograph.)

Figure 10.10 Russian MON100, front view. Back of MON100s can be seen in the background. (Author's photograph.)

- **Markings:** Tan, brown or green body with black markings is common. Other colors, stamped or stenciled markings and symbols may also be present.
- **Common fuze configurations:** Command initiation with hand-compressed "clackers" or tripwires set for pull or tension release.
- **General safety precautions** for directional fragmentation landmines include:
  - HE, frag, movement and B/T.
  - Safety precautions for the fuze if present.
  - EMR, static, and W/T if electrical components are seen or suspected.

Note: Do not cut tripwires.

   2. **Landmine, Anti-Tank (AT):** The landmine configurations covered under this group are designed to explode with enough force to disable or destroy tanks and other large vehicles. Depending on the configuration, explosive fillers can range from a little more than 1 lb to over 25 lb (0.5 to 11 kg). Fuze settings and types vary, but magnetic and seismic influence fuzing must always be considered. The most common fuze types require direct pressures of 150 to 400 lb (68 to 181 kg); however, some mines, such as the British Barmine (Figure 3.25 in Chapter 3), employ a hydraulic pressure fuze that allows them to differentiate between wheeled and tracked vehicles. When a wheeled vehicle

Figure 10.11 Czechoslovakian PT-MI-BA-III AT mine with a bakelite outer body. Note the fire and impact damage to the top of the mine, but the fuze remains intact and functional. (Author's photograph.)

drives over the mine, pressure is applied for a short duration, but the sustained pressure of a tracked vehicle will cause the mine to function.

2a. Landmine, AT, blast is a non-fragmentation-producing mine designed to target armored vehicles with explosive blast effects (Figure 10.11). Most designs are deployed just below the surface of the ground. When functioned by a vehicle driving over the mine, the explosive force follows the easiest path of resistance and focuses the energy upward away from the ground, thus maximizing the energetic potential against the target.

General identification features associated with AT, blast landmines include:

- **Appearance and materials:**
  - Deployed below the surface of the ground.
  - Metal or plastic body.
  - Multi-piece construction.
  - A primary fuze and multiple auxiliary fuze wells for boobytrapping.
  - A means of manual arming.

- **Markings:** Tan, brown or green body with black markings is common. Other colors, stamped or stenciled markings and symbols may also be present.

- **Common fuze configurations:** Direct pressure; magnetic and seismic influence. B/Ts and internal fuzing with S/D or A/D features are included in some designs.
- **General safety precautions** for AT, blast landmines include:
  - HE, frag, movement, B/T.
  - EMR, static, and W/T if electrical components are seen or suspected.
  - Safety precautions for the fuze if present.

2b. Landmine, AT, with shaped charge or Explosively Formed Projectile (EFP) is designed to destroy tanks and armored vehicles by penetrating the bottom of the hull with an EFP as seen in Figures 10.12 and 10.13 or a shaped charge. Most designs are deployed below the ground with a tilt-rod fuze to ensure the mine functions when the tank is above it (Figure 10.12). The AT version of the FASCAM mentioned in the introduction is the BLU-91/B, which sits on the surface when deployed. The BLU-91/B has a magnetic influence mine with an EFP capable of defeating the underside of a tank (Figure 10.13).

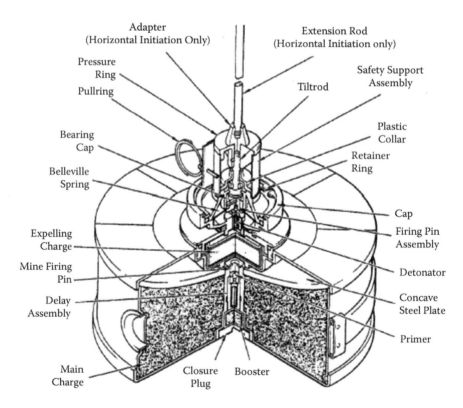

Figure 10.12 Line drawing of a U.S. M21 AT mine that incorporates an EFP design. (From U.S. military TM.)

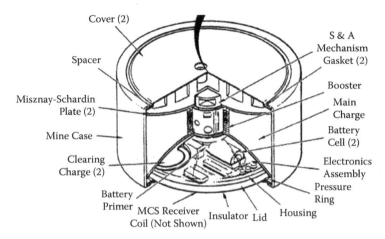

Figure 10.13 Line drawing of a U.S. BLU-92/B. APERS dispensed mine has four tripwires deploying from each side to ensure full area coverage. (From U.S. military TM.)

General identification features associated with AT landmines include:

- **Appearance and materials:**
  - Deployed below the surface or above ground.
  - Metal body (may have plastic components).
  - Multi-piece construction.
  - May have a tilt-rod.
  - The lack of an obvious fuze suggests an internal influence fuze.
  - A means of manual arming for tilt-rod type fuzes. Dispensed mines are usually electronically armed.

- **Markings:** Tan, brown or green body with black markings is common for hand placed mines. Dispensed mines may or may not be painted. Other colors, stamped or stenciled markings and symbols may also be present.
- **Common fuze configurations:** For a conventional design such as the M21 (Figure 10.12), an impact initiated tilt-rod fuze will ensure an underbelly detonation. For sophisticated designs such as the BLU-91/B, a magnetic influence fuze protected by an internal A/D and low voltage S/D feature may be present.
- **General safety precautions** for AT landmines include:
  - HE, frag, movement, jet, B/T.
  - Safety precautions for the fuze if present.
  - EMR, static, and W/T if electrical components are seen or suspected.

3. Landmine, practice is designed to replicate live mines in size and shape; many practice mines contain robust explosive or spotting charges, while others are hollow versions of the live munition.

General identification features associated with practice landmines include:

- **Appearance and materials:** The construction features associated with the landmine they are designed to imitate.
- **Markings:** The body may be painted blue or black and marked "practice" in white or painted colors inconsistent with common schemes. Other colors, stamped or stenciled markings, and symbols may be present.
- **Common fuze configurations** may or may not be present depending on landmine type.
- **General safety precautions** for practice landmines include:
  - Movement.
  - HE, frag, ejection when a spotting charge is present.
  - Safety precautions for the fuze if present.
  - Observing all applicable safety precautions for the live landmine until positive identification is made.

Note: Practice means "practice"—not "inert."

Closing

Landmines, especially dispensed mines may be difficult to recognize. However, they are inherently dangerous and designed to be initiated by their victim. As such, always consider a landmine to be in a hazardous condition until proven otherwise.

Ordnance Group—Chemical

11

There's no sense to this objection. It is considered a legitimate mode of warfare to fill shells with molten metal which scatters upon the enemy and produces the most frightful modes of death. Why a poison vapor which kills men without suffering is to be considered illegitimate is incomprehensible to me. However, no doubt in time chemistry will be used to lessen the suffering of combatants.

Sir Lyon Playfair, 1854

Prelude

Many of the categories covered in earlier chapters have a "chemical group." However, due to the unique threats associated with chemical munitions, they are afforded their own chapter as a group, rather than being spread throughout previous chapters.

When conducting research on munitions containing White Phosphorus (WP), burning smoke, riot control agents, and a number of other fillers such as Tri-Ethyl Aluminum (TEA) (Figures 7.13 and 7.14 in Chapter 7), they are often listed under chemical ordnance. For the purpose of this text, "chemical ordnance" includes munitions containing a chemical substance designed to kill, seriously injure, or completely incapacitate through physiological effects. This definition includes nerve, blister, blood, and choking agents, as well as incapacitating agents due to the unique physiological effects they cause.

Introduction

The wartime employment of chemicals as weapons is a thousand years older than the discovery of black powder. By the mid-1800s scientists were producing deadly chemical agents. Surprisingly, using deadly chemicals as weapons faced the same moral questions heard today. During the Crimean War, Sir Lyon Playfair requested permission to use cyanide-filled projectiles to break the siege of Sebastopol. The British War Office condemned the idea as "inhumane and as bad as poisoning the enemy's water supply." Sir Playfair's response, at the beginning of this chapter, implied that a painless death

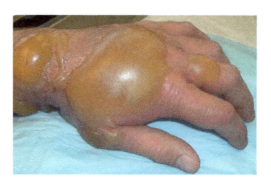

Figure 11.1 A person who came in contact with a Blister Agent. (DoD photograph).

results from exposure to deadly chemicals was not an accurate assessment (Figure 11.1.)

The first multinational attempt to limit the use of chemical weapons took place at the Brussels Convention in 1874, followed by conventions in 1899 and 1907. Unfortunately, all three conferences resulted in weak, vaguely worded resolutions. The result was that WWI became a large-scale chemical weapons testing ground.

World War I

Tactical deployment of chemical weapons reached its operational peak during WWI. It began with an attack by the Germans near Ypres, Belgium, on April 15, 1915, when 150 tons of chlorine gas was released from 6,000 cylinders and left to blow across the battlefield. Though somewhat successful, this method of delivery and dissemination was crude at best.

Over the next two years, phosgene and cyanide weapons were introduced and better delivery and dissemination methods were developed. On July 12, 1917, a German artillery attack delivered mustard agent in a far more efficient manner, resulting in 20,000 casualties. The full potential of a deadly chemical agent, efficiently delivered and disseminated in a manner that maximized the operational potential, was now realized. Throughout the remaining years of WWI, just about every country involved deployed chemical weapons in massive numbers.

From World War I to Today

After WWI the Geneva Convention of 1925 specifically addressed restricting the potential use of chemical weapons; as with previous conventions, there were many violations between WWI and WWII.

In the late 1930s, Dr. Gerhard Schrader, a German industrial chemist, synthesized the first nerve agent: tabun (GA). Two years later, he synthesized sarin (GB), an even more toxic nerve agent. The bar for chemical agents, in terms of lethality, had risen.

Throughout WWII, with thousands of tons of chemical weapons in inventory, there were only a small number of chemical weapon deployments.

After WWII, many countries continued chemical weapons programs. In 1993, the Chemical Weapons Convention Treaty was signed by 165 countries. The treaty prohibited manufacturing or stockpiling chemical weapons and many countries, including the United States and the Soviet Union, began destroying their stockpiles.

The Variables

The successful deployment of chemical weapons requires a few variables to be addressed. In order to be effective, a toxic chemical or incapacitating agent must be delivered in an appropriate concentration to provide the required dosage:

1. **A toxic chemical or incapacitating agent** can be in solid, liquid, or gas form. The physical properties of an agent determine the best tactical means of disseminating it.
2. **Concentration** addresses the strength and persistence of an agent.
3. **Dosage** is the amount of agent a person is exposed to via inhalation, ingestion, contact with the skin, or a combination of two or all three.
4. **Dissemination** Most chemical munitions use an explosive burster to break open the body of the munition and spread the chemical filler. All agents spread differently, freeze and boil at different temperatures, react differently to precipitation, and have different molecular weights. All agents can be greatly affected by the heat resulting from the burster detonating when the agent is initially dispersed and then by wind speed and temperature, vertical temperature gradients (VTGs), and precipitation, as well as the topography of the target location. All of these variables will have an enormous effect on the agent.

The Agents

Chemical agents are classified by the physiological effects they produce.

- Nerve agents.
- Blister agents.
- Blood agents.

- Choking agents.
- Incapacitating agents.

Nerve agents affect the central nervous system and may be inhaled, ingested, or absorbed through the skin. The systemic effects of nerve agents are caused by their ability to inhibit cholinesterase, which hydrolyzes the neurotransmitter acetylcholine, resulting in an accumulation of acetylcholine in nerve junctions throughout the body. Symptoms include pinpointing of the pupils, difficulty breathing, confusion, twitching and jerking of the muscles, and convulsions.

Examples of nerve agents include:

- Nonpersistent agents: GA (tabun), GB (sarin), and GD (soman).
- Persistent agent: VX.

Blister agents (vesicants) were originally developed during WWI. Blister agents burn and blister the skin and other parts of the body they contact, such as the lungs if inhaled (Figure 11.1). Blister agents tend to be nonvolatile and present a greater contact threat than vapors. Some of the most commonly encountered chemical munitions outside military control are WWI vintage projectiles filled with blister agents.

Examples of blister agents include:

- Sulfur mustards: H, HD, HT, HL, and HQ.
- Nitrogen mustards: HN-1, HN-2, and HN-3.
- Arsenicals L (Lewisite): ED, MD, PD.
- Urticant: CX.

Blood agents interfere with oxygen utilization at the cellular level. The effect is somewhat similar to that produced by exposure to high levels of carbon monoxide. These agents are highly volatile and dissipate quickly in outdoor environments. However, they work very well in enclosed space and AC is still used in gas chambers for capital punishment.

Examples of blood agents include:

- Hydrogen cyanide: AC.
- Cyanogen chloride: CK.

Choking agents severely damage lung tissue, causing the lungs to fill with fluid and resulting in "dry land drowning." Some of these agents are volatile and dissipate somewhat quickly, but some, such as CG, are heavier than air and can linger in low areas for many hours under light wind conditions.

Examples of choking agents include:

- Chlorine: CL.
- Phosgene: CG.
- Chloropicrin: PS.
- Diphosgene: DP.

Military incapacitating agents: Riot control agents are often referred to as "incapacitating" agents as they may blur a person's vision and cause a burning sensation for a short period of time, although a person is not rendered fully incapacitated. On the other hand, military-grade incapacitating agents alter or disrupt the functions of the central nervous system (CNS) with hallucinogenic symptoms similar to LSD. Depending on the dose, a person can be fully incapacitated for days.

Examples of some of the known incapacitating agents include:

- 3-quinuclidinyl benzilate (QNB) (NATO name is "BZ").
- d-lysergic acid diethylamide (LSD-25).
- Agent 15.
- Kolokol-1.
- Fentanyl and cannabinol derivatives.

Ordnance Categories with a Chemical Group

Whenever there is a threat or higher than normal possibility that chemical ordnance may be present, the first step in determining the group should be to rule out "chemical." An example of a higher than normal possibility would include ordnance discovered in an area known to have been a former military impact area or battlefield where chemical munitions were tested or used. Other examples would include discoveries where confirmed chemical ordnance recoveries have previously occurred whether on land or sea. Many chemical munitions were disposed of at sea and are dredged up by ships or pushed onto beaches during storms.

Most chemical fillers used with ordnance have a thin liquid or thick, jelly-like consistency. As such, chemical munitions are often configured similarly to munitions loaded with WP or other liquids. The most accurate means of differentiating a chemical munition is the color codes and markings (Figures 2.1–2.3 in Chapter 2). When confronted with an old, damaged, or repainted munition in which "chemical" cannot be ruled out as a possibility, assume the munition may contain chemical filler until proven otherwise.

Chemical weapon groups are presented in chapter order and include:

- Projectiles.
- Bombs.

Figure 11.2 An unfuzed 3 in. Stokes mortar, which was a common WWI chemical projectile. Stokes mortars are spin stabilized and do not have fins. (Author's photograph.)

- Rockets.
- Missiles.
- Submunitions.
- Landmines.

1. Projectile, chemical: The most common design involves a chemical agent sealed in the projectile by a burster adapter, which contains a high-explosive burster extending from the fuze well down the center of the projectile. Figure 4.52 in (Chapter 4) provides an example of this internal configuration, which is consistent with Figures 11.2 and 11.3. Additionally, Figures 4.9 and 4.51 offer examples of obvious and somewhat obscure booster adapters. The internal configuration of Figure 11.4 is substantially different from Figure 11.3, but the external identification feature, an adapter between the fuze and the projectile is present.

Upon fuze functioning, the burster detonates, breaking the projectile body into pieces while dispersing the chemical agent.

General identification features associated with chemical projectiles include:

- **Appearance and materials:**
  - A solid, one-piece body of robust construction with rotating band(s), gas check bands, or an obturator band (Figures 4.14–4.20).

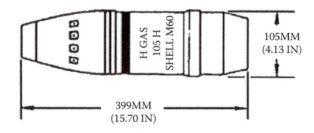

Figure 11.3 Line drawing of a common chemical projectile configuration with adapter and chemical marking "H GAS." (From U.S. military TM.)

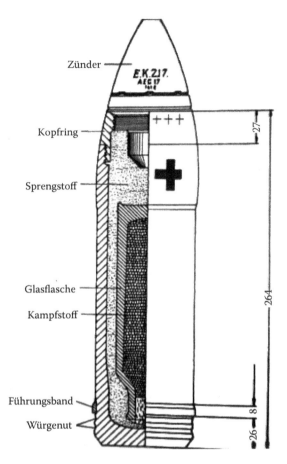

Figure 11.4 Line drawing of a German WWI era chemical projectile. Sprengstoff = explosive; Glasflasche = glass (bottle); Kampfstoff = chemical agent; Kopfring = stop ring (adapter). Though configured and constructed differently from that in Figure 11.3, the identification features are consistent. (From U.S. military TM.)

- A burster adapter is between the projectile body and fuze, which seals in the chemical agent.
- The adapter booster may have wrench flats or spanner holes.
- A solid base is common.
- If a tail boom or fins are present, there will not be a venturi consistent with a rocket or an open tube consistent with a rifle grenade at the base.

- **Markings:** A gray body with yellow, green, or red markings is common. Other colors, stamped or stenciled markings, and symbols may also be present.

- **Common fuze configurations:**
  - PD fuzing that functions upon impact with the ground.
  - MT or ET fuzing, which functions during flight.

- **General safety precautions** for chemical projectiles include:
  - High explosive (HE), fragmentation (frag), movement, and chemical.
  - Safety precautions for the fuze if present.

Note: Add ejection for base-ejecting models.

2. Bomb, Chemical: Chemical-filled bombs are often slightly modified high-explosive bombs and the modifications may be difficult to detect. Figure 11.5 offers an example of a basic configuration, which involves a chemical agent sealed in the bomb body with a high-explosive burster extending

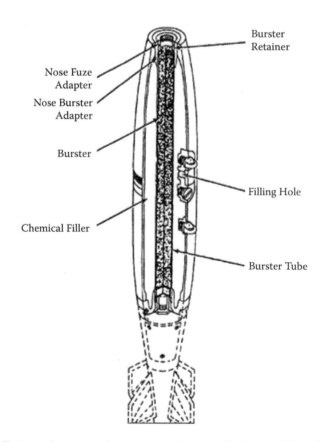

Figure 11.5 Line drawing of a U.S. MK94, 500 lb (226 kg) bomb containing 108 lb (49 kg) of GB and a burster with over 16 lb of high explosives. (From U.S. military TM.)

from the fuze well, down the center of the bomb. Upon fuze functioning, the burster detonates, breaking the bomb body into pieces while dispersing the chemical agent.

General information associated with chemical bombs includes:

- **Appearance and materials:**
  - Solid, one-piece body of robust construction.
  - Welded or fixed plugs over the fuze wells.

- **Markings:** A gray body with yellow, green, or red markings is common. Data plates may be present on the base. Other colors, stamped or stenciled markings, and symbols may also be present. Note the bands, which are yellow and green, between the lugs in Figure 11.5.
- **Common fuze configurations:**
  - Impact fuzing.
  - VT fuzing, which functions during flight.

- **General safety precautions** for chemical bombs include:
  - HE, frag, movement, chemical.
  - Safety precautions for the fuze if present.

3. Rocket, chemical: The most common design is similar to a chemical projectile in which the chemical agent is sealed in the rocket warhead by a burster adapter. The high-explosive burster extends from the fuze well, down the center of the warhead. Upon fuze functioning, the burster detonates, breaking the warhead body into pieces while dispersing the chemical agent. Figures 7.20 and 7.21 (Chapter 7) offer examples of nose and base booster adapter configurations on WP rocket warheads that may also indicate chemical agent filler.

General identification features associated with chemical rockets include:

- **Appearance and materials:**
  - Other than the adapter booster, the warhead is of similar shape and size as HE or HEAT warheads.
  - Adapter booster may have wrench flats or spanner holes.
  - There is a motor on the aft end with the warhead on the forward end.
  - Fin assembly is on the aft end of the motor.
  - One or more venturis are on the base of the motor.

- **Markings:** A gray body with yellow and red markings is common. A brown band may be present on the motor. Other colors, stamped or stenciled markings, and symbols may also be present.

- **Common fuze configurations:**
  - PD fuzing.
  - VT fuzing, which functions during flight.

- **General safety precautions** for chemical rockets include:
  - HE, frag, movement, chemical.
  - Electromagnetic radiation (EMR), static, and ejection for an unfired motor.
  - Safety precautions for the fuze if present.

4. Missile, Chemical: Missiles designed to deploy chemical agents tend to be of considerable size—as large as or larger than the SA-2 pictured in Figure 8.9 in Chapter 8. As the focus of this book is the smaller ordnance items more commonly encountered outside military control, these missiles will not be discussed.

5. Submunition, Chemical: Though uncommon, there are submunitions designed to deploy chemical agents. Many submunitions ignore conventional color-coding schemes, making positive identification more difficult.

6. Landmine, Chemical: Chemical-filled landmines are oftentimes slightly modified Anti-Tank (AT) or Anti-personnel (APERS) mines and the modifications may be difficult to detect. For example, the M23 VX landmine shown in Figures 11.6 and 11.7 has four raised projections that are not found on the M15 AT mine of similar design and shape; however, the pressure requirements for the fuze usually remain unchanged.

General identification features associated with chemical landmines include:

- **Appearance and materials:**
  - Deployed below the surface of the ground.
  - Metal or plastic body.
  - Multipiece construction.
  - A means of manual arming (removing a shipping cap may arm the mine).

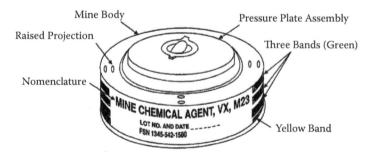

Figure 11.6 Line drawing of a U.S. M23 chemical (VX) mine. (From U.S. military TM.)

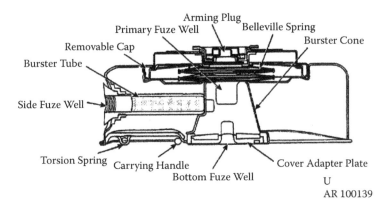

<figure>Arming Plug
Primary Fuze Well   Belleville Spring
Removable Cap   Burster Cone
Burster Tube
Side Fuze Well
Torsion Spring Carrying Handle   Cover Adapter Plate
Bottom Fuze Well
U
AR 100139</figure>

Figure 11.7 Line drawing of an M23, which contains over 10 lb (4.5 kg) of VX dispersed by a 14 oz (397 g) burster. (From U.S. military TM.)

- **Markings** Gray body with yellow, green or red markings is common. Other colors, stamped or stenciled markings, and symbols may also be present.
- **Common fuze configurations:** Direct pressure. Pressure plate may be visible, but fuzing is usually internal; self-destruct (S/D) or anti-disturbance (A/D) fuzing is included on some designs.

Note: In addition to the primary fuze, side and bottom fuze wells are common (Figure 11.7).

- **General safety precautions** for chemical landmines include:
  - HE, frag, movement, chemical, boobytrap (B/T).
  - EMR, static, and wait time (W/T) if electrical components are seen or suspected.

- **Markings:** Gray body with yellow, green or red markings is common. Other colors, stamped or stenciled markings, and symbols may also be present.

Closing

The injuries seen in Figure 11.1 were caused by a munition that was almost 100 years old. When encountered, chemical munitions pose a significant short and long-term threat to a larger area. Always consider a suspected chemical munition to be in a hazardous condition until proven otherwise.

Ordnance Category— 12
Underwater Ordnance

Damn the torpedoes, full speed ahead.

Admiral David Farragut, Battle of Mobile Bay, August 5, 1864

Introduction

Most underwater ordnance groups are designed to influence shipping by denying access or destroying an enemy's ships. During the US Civil War, Confederate forces did this by deploying "torpedoes"—what today are described as moored mines. At the beginning of the battle for Mobile Bay, Union forces attempting to navigate through a complex minefield lost the monitor *USS Tecumseh,* which struck a torpedo and immediately sank. For a moment it appeared as if the minefield would achieve its objective of denying the Union flotilla access to Mobile Bay. But Admiral Farragut's orders led to Union forces successfully passing through the minefield.

Some accounts, have the Admiral's actual words as "Damn the torpedoes! Four bells! Captain Drayton, go ahead! Jouett, full speed!"—which captures the intent. During the passing of the forts, additional ships were lost and there are accounts of sailors hearing torpedoes bang and scrape along their ships' hulls without exploding—technical shortfalls in the fuzing that would not be expected today.

Operating in an underwater environment is extremely different from land-based operations and this is reflected in the munitions designed to support these operations. The designs and functioning methods of underwater munitions are so different that US Navy EOD (Explosive Ordnance Disposal) technicians have to attend an additional two months of training focused solely on this ordnance category. Many of these munitions are commonly found washed up on beaches around the world, oftentimes obscured by sea growth.

Underwater munitions are designed to damage, disable, or sink ships; attack combat swimmers; mark targets; locate survivors; determine wind speed or direction; and a host of other tasks. This chapter will introduce the ordnance groups classified as underwater ordnance.

The Seven-Step Practical Process
Applied to Underwater Ordnance

Examples of different designs, features, color codes, markings, and construction features are provided throughout this chapter.

Step 1: Approach and initial interrogation. Attempt to identify a munition at a distance with the use of binoculars. If an approach is made, avoid all venturis, fuze-sensing elements, and propellers. Armed and active or damaged, sensing elements may "see" a person approaching, consider the person a valid target, and function as designed. Propellers are also capable of self-activation and must be avoided.

On some underwater ordnance groups, it is common to find detailed information stamped into the body, including the model, weight, type and amount of explosives, date of manufacture, and a shipping address. If present, use this information to research the munition. If not, measurements of the major diameter at its widest point and the overall length of the munition will help identify it. If possible, make note of any identifiable features. All findings, including measurements, color codes, markings, key identifying features, and any possible damage, are documented and the munition is photographed.

In addition to the overall configuration, there are three features that will greatly assist in answering steps 3, 5, and 7:

1. The diameter.
2. The overall length.
3. Presence of propellers, fins (fixed and movable), and wires protruding from the body.

Step 2: Determine fuze/exploder type and condition: Some underwater ordnance uses the term "exploder" versus "fuze" to describe what has thus far been generally described as "the fuzing" for an ordnance item.

If a munition has been deployed, the fuze/exploder is considered to be armed (step 5). If a fuze/exploder is damaged, pins have been removed, or any alterations have been made to the munition, it is considered armed. If visible, measurements for the fuze/exploder are taken separately from the munition.

Note: If deployed or discovered in or near water, assume all underwater ordnance is fuzed.

Step 3: Determine ordnance category. This category covers all ordnance designed to be deployed underwater.

Step 4: Determine ordnance group. Identifying characteristics associated with each group will be mentioned throughout this chapter.

Step 5: Determine if the munition was deployed. Inspect the munition for impact-related damage and missing pins or clips. If found in or near water, assume the munition was deployed.

Step 6: Determine safety precautions that apply to the munition. Safety precautions for the groups are covered in this chapter. Chapter 3 addresses the safety precautions associated with various fuzes.

Note: Adhere to all safety precautions that apply.

Step 7: Identify the munition. Apply the totality of all construction characteristics and other identifying features to determine the group to which an ordnance item belongs and, if feasible, positively identify the munition and all possible fuzing configurations.

Groups

A basic definition of underwater ordnance is a munition designed to be deployed in water. It does not apply to other ordnance categories that may have been dumped, fired into, or otherwise come to be underwater. In order to provide a coherent flow, the underwater ordnance category is divided into the following primary and supplemental groups:

1. Torpedoes.
2. Anti-submarine.
   a. Depth bombs.
   b. Depth charges.
   c. Hedgehogs.

3. Mines.
   a. Moored.
   b. Bottom.
   c. Limpet.
   d. Shallow water.

4. Sound signals.
5. Pyrotechnic markers.

1. Torpedoes: Are self-propelled weapons designed to sink surface ships and submarines with a high explosive warhead ranging from 100 to over 800 lb (45 to 362 kg). Torpedoes can be tube-fired from ships and submarines, dropped from aircraft and tracked to their target via wire guidance, or perform active and passive acoustic homing (Figures 12.1 and 12.2). "Torpedoes" of the American Civil War were not self-propelled and today would be grouped as moored mines (Section 3a). General identification features associated with torpedoes include:

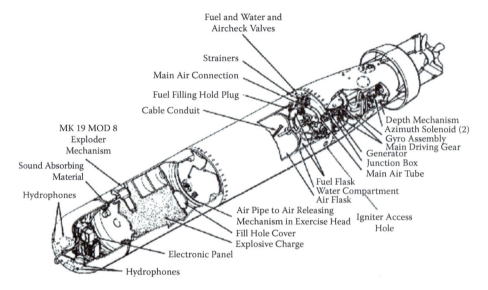

Figure 12.1 Example of a torpedo configuration (there are many different designs). (From U.S. military TM.)

Figure 12.2 The Howell torpedo. The first (successful) torpedo developed by the United States was created by Lt. Commander John Howell in 1870. On display at the Naval Undersea Museum, Keyport WA. (Photograph from the U.S. Naval War College Museum website.) In March 2013, a Howell torpedo was recovered by a Navy EOD team off the coast of San Diego, CA.

- **Appearance and materials:**
  - Multi-piece construction.
  - Over 20 in. in diameter (U.S. torpedoes are 21 in.).
  - May be over 20 ft in length.
  - May have single or multiple screws (propellers) on the aft end.
  - Vertical fins toward the aft end with a movable surface for direction changes.
  - A shroud encircling the fins to protect the screws.
  - If wire guided, wire possibly attached at the aft end.
  - If deployed from aircraft, lugs or cleats may be present.

- **Markings:** Identification information may be stamped on the outside of each section. Other colors, stenciled markings, and symbols may also be present.
- **Common fuze or exploder configurations:** Impact, influence (magnetic and acoustic); as torpedo fuzing is internal, always assume a fully operational exploder is present until proven otherwise.
- **General safety precautions** for torpedoes include:
  - HE, frag, movement.
  - Ejection if marker ejection ports are present.
  - Chemical for fuel.
  - Electromagnetic radiation (EMR), static, influence, and wait time (W/T) for fuzing.

Note: Torpedoes contain a number of hazards, including toxic fuels, corrosive substances, high-voltage components, high-pressure gas cylinders, and explosive actuators.

Note: Do not remove encrusted marine growth from an ordnance item.

2a. Depth bombs: Are designed to destroy submarines and other submerged targets with a high-explosive main charge ranging from 200 to over 500 lb (90 to 226 kg). Delivered from aircraft in a similar manner to that of air-dropped bombs, they can also be deployed against land targets.

General identification features associated with depth bombs include (Figure 12.3):

- **Appearance and materials:**
  - Multi-piece construction.
  - Fins.
  - Lugs or cleats.
  - May have four fuze wells (nose, base, and two transverse).

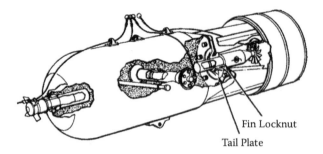

Fin Locknut

Tail Plate

Figure 12.3 Line drawing of a depth bomb with nose, tail, and transverse fuzing. (From U.S. military TM.)

- **Markings:** Identification information may be stamped between the lugs or cleats. Other colors, stenciled markings, and symbols may also be present.
- **Common fuze configurations:** Hydrostatic or impact fuzing for waterborne deployment, impact for use against land-based targets. Fuze wells may be in the nose, tail, or side (transverse). Hydrostatic fuzing will function at a preset depth.
- **General safety precautions** for depth bombs include:
  - HE, frag, movement.
  - EMR, static, influence, and W/T for fuzing.

Note: Do not remove encrusted marine growth from an ordnance item.

2b. Depth charges: Are deployed from ships via rails or projected by a special launching platform to destroy submarines. Smaller, hand-deployed depth charges are used for other submerged targets such as combat swimmers. The high-explosive main charges vary from 3 to 300 lb (1.36 to 136 kg).

General identification features associated with depth charges include (Figures 12.4 and 12.5):

- **Appearance and materials:**
  - Multi-piece construction.
  - Fins.
  - May have nose, base, or transverse fuze wells.

- **Markings:** Identification information is usually stamped on the body. Other colors, stamped or stenciled markings, and symbols may also be present.
- **Common fuze configurations:** Ship Launched: Hydrostatic arming with hydrostatic, magnetic, acoustic, or impact fuzing, usually located in the nose or base. Hand Deployed: Direct arming (similar

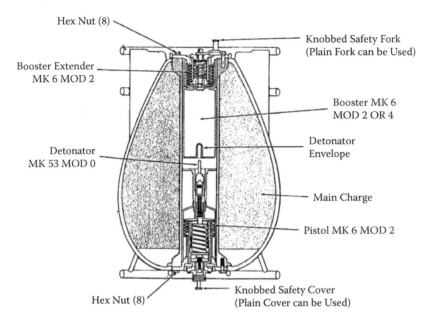

Hex Nut (8)

Booster Extender
MK 6 MOD 2

Detonator
MK 53 MOD 0

Knobbed Safety Fork
(Plain Fork can be Used)

Booster MK 6
MOD 2 OR 4

Detonator
Envelope

Main Charge

Pistol MK 6 MOD 2

Hex Nut (8)

Knobbed Safety Cover
(Plain Cover can be Used)

Figure 12.4 Line drawing of an MK 9 MOD 4, which contains a 200 lb (91 kg) main charge and 40 lb (18 kg) of lead to increase its rate of descent. (From U.S. military TM.)

to hand grenade) and hydrostatically fired. Hydrostatic fuzing will function at a preset depth.

- **General safety precautions** for depth charges include:
  - HE, frag, movement.
  - EMR, static, influence, and W/T for fuzing.

Note: Do not remove encrusted marine growth from an ordnance item.

2c. Hedgehogs: Are fired as a short-range rocket from the front of a ship, one at a time, or in salvos of up to 24 at a time, to destroy submarines and other submerged targets. Deployed in conjunction with depth charges, the high-explosive main charge for hedgehogs ranges from 36 to 200 lb (16 to 90 kg).

General identification features associated with hedgehogs include (Figures 12.6 and 12.7):

- **Appearance and materials:**
  - Cylindrical or teardrop shape.
  - Rocket motor configuration with venturi in the base.
  - Fins.
  - Nose fuze.

Figure 12.5 MK 9 MOD 2 depth charge. (Photograph courtesy of Tom Conte, U.S.N, Ret.)

- **Markings:** Identification information may be stamped on the body. Other colors, stenciled markings, and symbols may also be present.
- **Common fuze configurations:** Hydrostatic arming with hydrostatic, magnetic, acoustic, or impact fuzing in the nose. Hydrostatic fuzing will function at a preset depth.
- **General safety precautions** for hedgehogs include:
    - HE, frag, movement, ejection, EMR, static (for warhead and rocket motor).
    - EMR, static, influence, and W/T for fuzing.

Note: Do not remove encrusted marine growth from an ordnance item.

3. Mines: As with landmines (Chapter 10), underwater mines are specifically designed to self-destruct or be functioned by their intended victim and

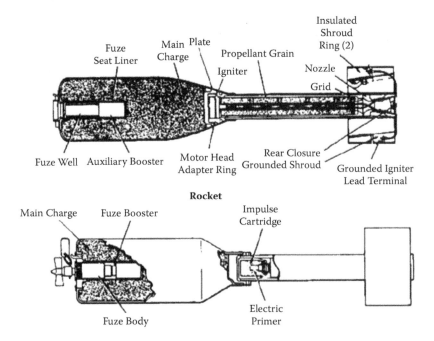

Figure 12.6 Line drawing of the internal configuration of the 7.2 in. MK10 and MK11 Hedgehog. The Navy did not classify Hedgehogs as "rockets" because the rocket motor was expended before the Hedgehog left the launch platform. (From U.S. military TM.)

should be considered boobytrapped. They also share the same problems when hostilities end as poor record keeping and the underwater environment make locating some types of mines very difficult. For example, in 2009 during an annual NATO naval exercise off the coast of Estonia, 64 underwater mines from WWI and WWII were discovered. Since 1994, over 600 mines have been located during NATO minesweeping exercises off the Estonian coast.

The types of mines covered include:

- Moored and drifting.
- Bottom.
- Limpet.
- Shallow water.

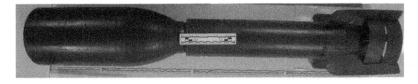

Figure 12.7 A ground-fired T-37 demolition rocket adapted from the 7.2 in. Hedgehog design. (Author's photograph.)

3a. Moored and drifting mines: Are most commonly deployed from ships with a cart that acts as an anchor; these mines are designed to destroy ships and submarines. Containing high-explosive main charges of about 200 lb (90 kg), much of the mine body is empty to make it buoyant. Moored mines are designed to remain anchored in place (Figure 12.8), whereas drifting mines are equipped with a lighter anchor that will fix the mine at a certain depth, but allow it to drift with the current. Both moored and drifting mines operate in the same way and employ similar fuzing. Both types can also become detached from their mooring and float on the surface. Since there is the possibility of these mines eventually separating from their mooring and becoming a threat to all shipping, many designs include a flooder assembly and sterilizer (Figure 12.9) to sink or render the mine inoperable.

General identification features associated with moored and drifting mines include:

- **Appearance and materials:**
  - Round, oval, or cylindrically shaped, 3 to 4 ft in diameter.
  - Steel body.
  - "Horns" protruding between 4 and 10 in.
  - May have mooring chain or cable attached to base.

- **Markings:** Identification information may be stamped on the body. Other colors, stenciled markings, and symbols may also be present.

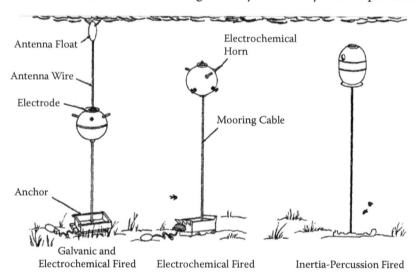

Figure 12.8 Line drawing that depicts three deployment variations of moored mines anchored to the sea floor after successful deployment. Note the antenna on the galvanic and electromechanical fired mine. In this configuration, antennas are used to increase the vertical contact area of the mine. (From U.S. military TM.)

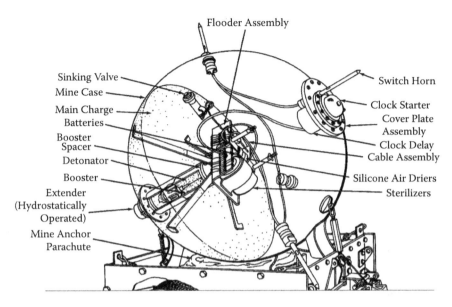

Figure 12.9 Line drawing that shows the internal configuration of a moored mine, fixed in its predeployment configuration on the cart that will act as its anchor. (From U.S. military TM.)

- **Common fuze configurations:** Impact, influence (magnetic and acoustic) fuzing incorporating chemical horns, switch horns and antennas (Figures 12.10). Internal flooder assembly and sterilizers (Figure 12.9) designed to sink or render a mine inoperable may also be present.
- **General safety precautions** for moored and drifting mines include:
  - HE, frag, movement, boobytrap (B/T).
  - EMR, static, influence, and W/T for fuzing.

Note: Do not remove encrusted marine growth from an ordnance item.

Note: Additional hazards such as explosive bolts, explosive cable cutters, explosive flooders, and sterilizers are present in or on many mines.

3b. Bottom mines: As the name implies, bottom mines are deployed from ships, submarines, and aircraft; they sink and remain on the bottom (Figure 12.11). Designed to destroy submarines by functioning close enough to the vessel to crush its hull, bottom mines apply a different strategy when targeting surface ships. To do this, bottom mines explode and produce an air bubble as a ship passes over. As the air bubble rises, water pressure decreases, and the air bubble increases in size (per Boyle's law). With high-explosive main charges that are oftentimes well over 1000 lb (450 kg), upon reaching the surface, the air bubble is large enough for the targeted ship to

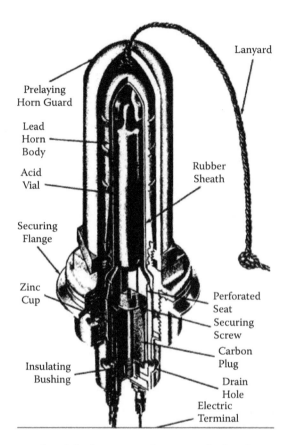

Prelaying
Horn Guard

Lead
Horn
Body

Acid
Vial

Securing
Flange

Zinc
Cup

Insulating
Bushing

Lanyard

Rubber
Sheath

Perforated
Seat

Securing
Screw

Carbon
Plug

Drain
Hole

Electric
Terminal

Figure 12.10 Moored and drifting mine fuzing includes chemical and electrical "horns" or protrusions that function the mine on contact. This example of an electrochemical horn, as seen deployed in Figure 12.8, consists of a lead or brass sheath containing a glass vial filled with electrolyte. The vial is connected to a wet-cell battery, which is connected to the mine firing circuit. When the horn is broken by contact with a target, the vial breaks, filling the battery with electrolyte, which immediately produces current flow to the detonator, functioning the mine. (From U.S. military TM.) Switch horns, as seen in Figure 12.9, are simple electric switches. When the horn is broken by contact with a target, the firing circuit closes and current flows to the detonator, functioning the mine.

dip or drop into. The sudden shift in support along the main axis (keel) of the ship literally snaps the hull.

General identification features associated with bottom mines include:

- **Appearance and materials:**
  - Cylindrically shaped mines possibly over 20 in. (51 cm) in diameter and over 71 in. (180 cm) in length.
  - Dome-shaped mines possibly over 48 in. (122 cm) in diameter.

Figure 12.11 MK 52 bottom mine, which incorporates a range of 1-hour to 90-day arming delay, a 30-count ship counter, an explosive sterilizer, and a hydrostatic arming device. Containing 625 lb (283 kg) of HBX-1 (see Appendix D), it can employ a pressure, acoustic, or magnetic fuze firing mechanism. (Photograph courtesy of Tom Conte, U.S.N. Ret.)

- Steel body.
- May have an outer plastic-like material to conceal the mine.
- Lugs or cleats if air dropped (Figure 12.11).

- **Markings:** Identification information may be stamped on the body. Other colors, stenciled markings, and symbols may also be present.
- **Common fuze configurations:** Transverse influence (magnetic and acoustic) fuzing incorporating ship counters, fuze-arming delay mechanisms ranging from 1 hour to many months, clock mechanisms, hydrostatic arming devices, and flooder or sterilizer assemblies designed to render a mine inoperable.
- **General safety precautions** for bottom mines include:
  - HE, frag, movement, B/T.
  - EMR, static, influence and W/T for fuzing.

Note: Do not remove encrusted marine growth from an ordnance item.

Note: Additional hazards such as explosive bolts, explosive flooders, and sterilizers are present in or on many mines.

3c. Limpet mines: The name "limpet" is derived from the resemblance these mines have to the shell of a mollusk. Limpets are small mines placed on ships below the waterline by combat swimmers and designed to look as if they are part of the ship (Figure 12.12). Secured in place with strong magnets, epoxy, or suction cups, they contain high-explosive main charges of 2 to 5 lb (0.9 to 2.3 kg), but with the water acting as a tamp, the energy of these small

Figure 12.12 A U.S. limpet mine. The mechanical time fuze threads into the protrusion on the far side. The anti-withdrawal device is armed by pulling the ring at the bottom and removing the wire. (Author's photograph.)

charges is maximized. Limpet mines are placed near critical areas of a ship so as to damage or disable versus sink a vessel.

General identification features associated with limpet mines include:

- **Appearance and materials:**
  - A flat side that goes against a target.
  - Magnets, suction cups, or other means of affixing to a target.
  - A round or cylindrical dome shape, approximately 12 to 18 in. (340mm to 457mm) in diameter or length.
  - Steel or aluminum body.
  - May have an outer plastic-like material for concealment.

- **Markings:** Identification information may be stamped on the body. Other colors, stenciled markings, and symbols may also be present.
  - **Common fuze configurations:** Mechanical time fuze as a primary means of functioning and an anti-lift or anti-withdrawal device that is armed after the mine is placed.
- **General safety precautions** for limpet mines include:
  - HE, frag, movement, B/T.
  - EMR, static, influence, and W/T for fuzing.

Note: Do not remove encrusted marine growth from an ordnance item.

Figure 12.13 Three different configurations of shallow water mines. (Author's photograph.)

3d. Shallow-water mines: Are deployed in the shallow waters of rivers, swamps, and the surf zone of lakes and oceans. Early designs were little more than anti-tank (AT) mines weighted down or anchored with stakes, but the violent environment of a surf zone or swift river currents resulted in fuzing malfunctions. The shapes, sizes, and configurations of newer designs vary greatly (Figure 12.13), but they are constructed and fuzed to operate efficiently in these environments. Containing high-explosive main charges of about 5 to 30 lb (2.3 to 13.6 kg), shallow-water mines usually target landing craft and small boats operating in waters less than 20 ft (6 m) in depth.

General identification features associated with shallow-water mines include:

- **Appearance and materials:**
  - Body shapes and sizes varying greatly.
  - Holes, threaded adapters, or other means of anchoring the mine in place.
  - Steel or aluminum body.
  - May have an outer plastic-like material for concealment.

- **Markings:** Identification information may be stamped on the body. Other colors, stenciled markings and symbols may also be present.
- **Common fuze configurations:** Impact and, to a lesser extent, influence fuzing with anti-lift, anti-withdrawal, or other B/T devices.
- **General safety precautions** for shallow-water mines include:
  - HE, frag, movement, B/T.
  - EMR, static, influence, and W/T for fuzing.

Note: Do not remove encrusted marine growth from an ordnance item.

4. Sound signals: Sound Underwater Signals (SUS) and Sound Fixing and Ranging (SOFAR) devices are designed to be deployed from aircraft, surface ships, and submarines. These small devices contain high-explosive main charges ranging from 0.25 to over 8 lb (0.11 to 3.7 kg). SUS and SOFAR devices are used for signaling, calibrating oceanographic equipment, and locating submarines and other submerged objects by tracking reflected sound waves.

General identification features associated with SUS and SOFAR devices include:

- **Appearance and materials:**
  - Multi-piece construction.
  - Made of steel or aluminum.
  - Usually between 14 and 22 in. (355mm and 559mm) in length.
  - Usually between 2 and 3 in. (51mm and 76mm) in diameter.
  - May have fins on the aft end.

- **Markings:** Identification information may be stamped on the device. Other colors, stenciled markings, and symbols may also be present.
- **Common fuze configurations:** Hydrostatic arming with hydro-static initiated fuzing to function at a preset depth. Some fuzes, such as the one shown in Figure 12.14, incorporate a cocked firing pin, or Cocked Striker (C/S).

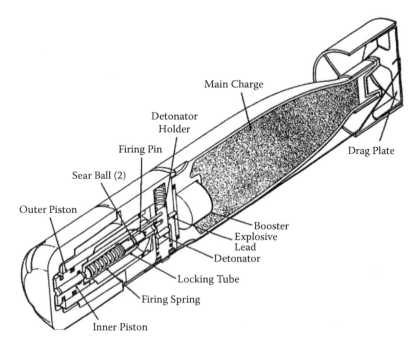

Figure 12.14 Line drawing of the internal configuration of a sound underwater signal (SUS) device. (From U.S. military TM.)

- **General safety precautions** for SUS and SOFAR devices include:
  - HE, frag, movement.
  - C/S for fuzing.

Note: Do not remove encrusted marine growth from an ordnance item.

5. Pyrotechnic markers: Are pyrotechnic devices used for signaling purposes or as floating markers, as they emit smoke for daytime and flame for nighttime deployment. Designed to be deployed from aircraft, surface ships, and submarines, these devices can contain Red Phosphorus (RP), pyrotechnic mixtures, or chemicals that react with water to produce fire or smoke. Smoke- or fire-producing materials range from 0.5 to over 5 lb (0.23 to 2.3 kg) and provide burn times of 10 to 30 minutes.

General identification features associated with pyrotechnic markers include:

- **Appearance and materials:**
  - Multi-piece construction.
  - Made of steel, aluminum, or wood.
  - Sizes vary, but the MK25 Mod 4 (Figure 12.15) is 18.5 × 2.9 in. (470mm × 73.7mm).
  - One end may have one or multiple "plugs" that will blow out to allow smoke or flame to exit.

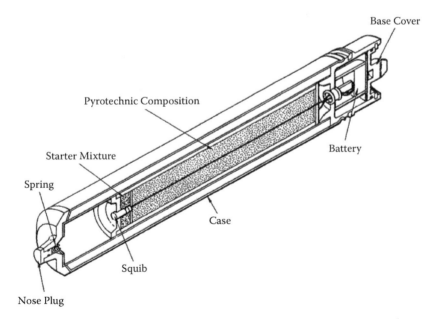

Figure 12.15 Line drawing of the internal configuration MK25 Mod 4 pyrotechnic marker. (From U.S. military TM.)

- **Markings:** Identification information may be stamped on the device. Other colors, stenciled markings, and symbols may also be present.
- **Common fuze configurations:** Some are armed upon deployment, but the most common configuration is soluble plugs that dissolve upon entering water, allowing seawater to fill and activate a battery, which electrically functions the munition.
- **General safety precautions** for pyrotechnic markers devices include:
  - Movement, fire.
  - White phosphorus if the marker contains RP.
  - Chemical if candle is burning.

Note: Do not look directly at a burning marker.

Note: Do not remove encrusted marine growth from an ordnance item.

Closing

Ordnance is commonly found on beaches, fishing nets and dredged off the sea floor. These munitions are designed to survive in harsh environments, evidenced by the intact Howell torpedo recovered in 2013, about 140 years after it was deployed. Do not be fooled by decayed outer appearance of an ordnance item that has been in an ocean or river, the internal components may be completely intact and pose a considerable threat.

Closing

13

When discussing military ordnance, no statement is more accurate than that of Francois Voltaire:

> The more I read, the more I acquire, the more certain I am that I know nothing.

François-Marie Arouet de Voltaire

Chapter 1 provided an overview of the energetics utilized in military ordnance. Chapter 2 offers a scientific method in the form of a practical deductive process that can be applied to interrogating an unknown ordnance item. Chapters 3 through 12 provide information on how ordnance is classified by the Category, Group, and sometimes the Type to which they belong. However, it is important to keep two things in mind. First, every ordnance group has a movement safety precaution and should only be moved by appropriately trained personnel after the munition has been identified. Secondly, though larger ordnance items were briefly covered, the focus of this text is primarily on the smaller ordnance items that are more commonly recovered outside of military control.

It is important to keep these limitations in mind as it is impossible to cover all military ordnance in one library, much less a single text, and there are many explosive devices utilized by the military that are not covered. For example, rare ordnance configurations such as the High-Explosive (HE) grenades with porcelain bodies manufactured by Japan during the latter part of WWII (Figure 13.1) may not be easily recognized. Other configurations, such as Figure 13.2, may include components from different categories and groups, put together in a haphazard manner. Though classified as ordnance, properly categorizing and grouping these munitions in order to apply the appropriate safety precautions may be difficult or impossible.

The explosive devices used to blow off or shatter the canopy of a fighter jet are not covered. Nor are the explosive components and rockets used to separate and eject a pilot from an aircraft, and deploy his or her parachute covered, as they are not classified as ordnance, but rather escape systems. The chaff and flares that are explosively ejected and explode or burn to attract guided missiles are not covered as they are classified as countermeasures.

Figure 13.1 Japanese WWII era porcelain hand grenade. (Photograph courtesy of Dan Evers.)

Figure 13.2 What appears to be a 60mm mortar body with a plugged fuze well and a modified base allowing the introduction of a hand grenade fuze. (Photograph courtesy of Didzis Jurcins.)

Figure 13.3 M110 gunflash simulator. (Author's photograph.)

There are also training simulators, which are pyrotechnic devices designed to simulate battlefield noises and the concussive effects (Figures 13.3 and 13.4). Classified as "special fireworks," land mine, grenade, and artillery simulators are usually constructed of cardboard or plastic. Most contain photoflash powder and are designed to produce a strong concussive effect or whistling sound prior to exploding.

There is a reason for the inclusion of movement safety precautions in every ordnance category and group. When applying the seven-step practical process to ordnance, including modified or unique munitions such as those seen in Figures 13.1 and 13.2, they are not to be moved. A munition is never touched or picked up unless the person doing so knows exactly what it is, how it works, and its condition, and has a reason or responsibility for taking such action. All of this will be determined at the conclusion of step 7.

Military ordnance is discovered every day by beach goers, hikers, police officers, firemen, history enthusiasts, archeologists, public utility personnel,

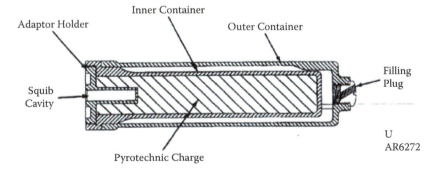

Figure 13.4 Line drawing of an M110 gunflash simulator. By design, the filling plug is removed and the void surrounding the substantial pyrotechnic charge is filled with fuel. (From U.S. military TM.)

and TSA personnel manning security checkpoints. In order to ensure public safety, as well as your own safety, consider applying the process laid out in Chapter 2 when you encounter any unknown item that appears to be ordnance, a component of an ordnance item, containing markings that suggest it is designed for military use.

Appendix A: Logic Trees

Logic Tree 1—Methods of Deployment

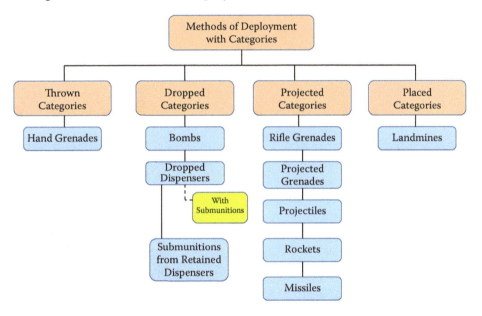

Logic Tree 2—Categories and Groups

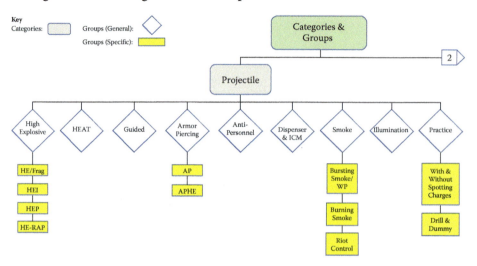

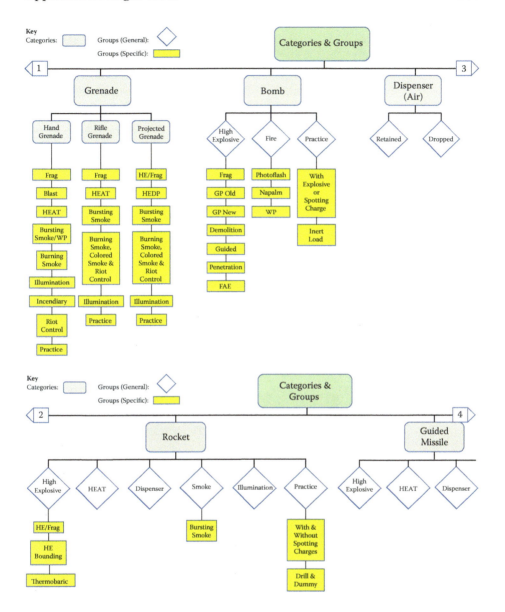

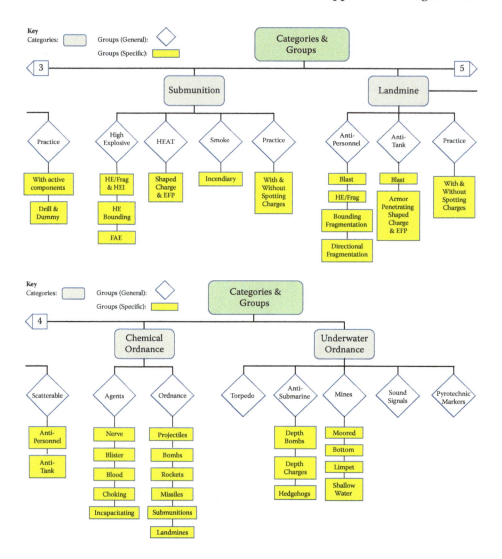

Logic Tree 3—Fuzes

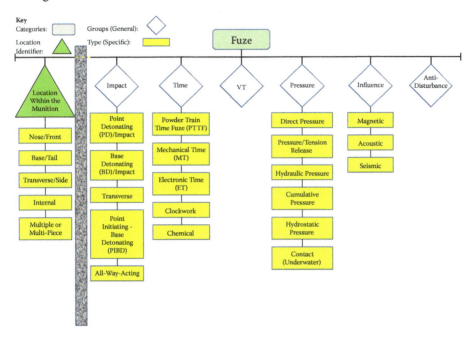

Logic Tree 4—Safety Precautions

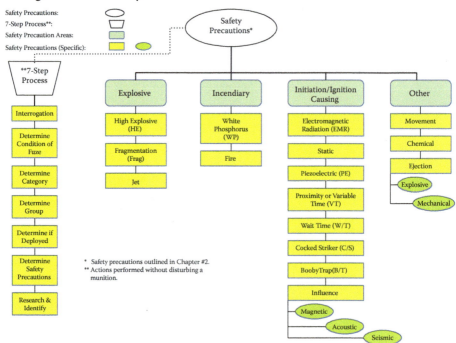

Safety Precautions:
7-Step Process**:
Safety Precaution Areas:
Safety Precautions (Specific):

**7-Step Process

- Interrogation
- Determine Condition of Fuze
- Determine Category
- Determine Group
- Determine if Deployed
- Determine Safety Precautions
- Research & Identify

Safety Precautions*

Explosive
- High Explosive (HE)
- Fragmentation (Frag)
- Jet

Incendiary
- White Phosphorus (WP)
- Fire

Initiation/Ignition Causing
- Electromagnetic Radiation (EMR)
- Static
- Piezoelectric (PE)
- Proximity or Variable Time (VT)
- Wait Time (W/T)
- Cocked Striker (C/S)
- BoobyTrap(B/T)
- Influence
- Magnetic
- Acoustic
- Seismic

Other
- Movement
- Chemical
- Ejection
- Explosive
- Mechanical

* Safety precautions outlined in Chapter #2.
** Actions performed without disturbing a munition.

Appendix B: Common Ordnance Related Abbreviations

AA: Anti-Aircraft.

AAA: Anti-Aircraft Artillery.

AC: Hydrogen Cyanide.

ACM: Anti-Armor Cluster Munition.

ADAM: Area Denial Artillery Munition.

AGM: Air-to-Ground Missile.

AIM: Air Intercept Missile.

AN: Army/Navy, used to designate munitions approved for use by the Army and the Navy.

AP: Armor Piercing.

APC: Armor Piercing Capped.

APHE: Armor-Piercing High Explosive. An AP munition with an HE charge in the base.

APAM: Anti-Personnel Anti-Material munition.

APDS: Armor Piercing Discarding Sabot.

APERS: Anti-Personnel.

APHEI: Armor Piercing High Explosive Incendiary.

AT: Anti-Tank.

BD: Base Detonating or Bomb Disposal.

BDU: Bomb Dummy Unit.

BE: Base Ejecting.

BLU: Bomb Live Unit.

B/T: Boobytrap.

CAD & PAD: Cartridge Actuated Device & Propellant Actuated Device.

CBU: Cluster Bomb Unit.

CG: Phosgene, a chemical agent.

CN: Chloroacetophenone, a tactical riot control agent.

CS: Named 2-chlorobenzalmalononitrile or o-chlorobenzylidene malono-nitrile, tactical riot control agents.

EMR: Electromagnetic Radiation.

EO: Explosive Ordnance.

EOR: Explosive Ordnance Reconnaissance.

ERW: Explosive Remnants of War.

ET: Electronic Time.

FAE: Fuel Air Explosive.

FASCAM: Family of Scatterable Mines.

FM: Titanium Tetrachloride.

FMU: Fuze Munition Unit.

Frag: Fragmentation.

FS: Sulphur Trioxidechlorosulfonic Acid Solution (white smoke).

GA: A nonpersistent nerve agent.

GB: A nonpersistent nerve agent.

GP: General Purpose.

GPLD: General Purpose Low Drag.

GPHD: General Purpose High Drag.

GPOS: General Purpose Old Style.

GPNS: General Purpose New Style.

H: A mustard agent.

HC: Hexachlorethane (smoke).

HD: A distilled mustard agent.

HE: High Explosive.

HEAT: High Explosive Anti-Tank.

HEDP: High Explosive Dual Purpose.

HEI: High Explosive Incendiary.

HEP: High Explosive Plastic (known as a "Squashhead" in some countries).

HERA: High Explosive Rocket Assist (see RAP).

HC: Hexachloroethane-zinc (White smoke mixture).

HVAP: High Velocity Armor Piercing.

HVAPDS: High Velocity Armor Piercing Discarding Sabot.

Illum: Illumination.

ICM: Improved Conventional Munition.

JATO: Jet Assist Take Off.

LE: Low Explosive.

LUU: Illumination Unit.

M: Model (designates model number on Army munitions).

MANPADS: Man Portable Air Defense System.

MK: Mark (designates model number on Naval munitions).

MOD: Modified or modification.

MSDS: Material Safety Data Sheet.

MT: Mechanical Time.

MTSQ: Mechanical Time Superquick.

PD: Point Detonating.

PDSD: Point Detonating Self-Destruct.

PDSQ: Point Detonating Superquick.

PI: Point Initiating.

PIBD: Point Initiating Base Detonating.
Projo: Projectile.
Prox: Proximity (interchangeable with VT).
PTTF: Powder Train Time Fuze.
PUCA: Pickup and Carry Away.
PWP: Plasticized White Phosphorus.
RAP: Rocket Assisted Projectile (see HERA).
RAAM: Remote Anti-Armor Mine.
RP: Red Phosphorus.
RPM: Revolutions Per Minute.
RSP: Render Safe Procedure.
SAM: Surface to Air Missile.
SAP: Semi-Armor Piercing.
SAPHE: Semi-Armor Piercing High Explosive.
SAPHEI: Semi-Armor Piercing High Explosive Incendiary.
SD: Self-Destruct.
SQ: Superquick.
S&A: Safe and Arm Device.
SVD: Stable Velocity of Detonation.
T: Tracer.
TDD: Target Detection Device (component of VT or Influence fuzing).
TP: Target Practice.
TSQ: Time Super-Quick.
UXO: Unexploded Ordnance. A deployed munition that failed to function as designed.
VD: Velocity of Detonation.
VT: Variable Time (Interchangeable with Proximity).
VX: A persistent nerve agent.
WP: White Phosphorus.

Appendix C: Functional Definitions for Ordnance-Related Terms

Adapter: A component used to facilitate a connection.

Adapter Booster: An adapter containing an explosive booster.

Anti-material: The addition of materials such as magnesium or zirconium to produce effects that can damage or destroy materials.

Approach and Initial Interrogation: The initial approach, observation, examination, and documentation of a recovered munition or munitions. Step 1 of the Seven-Step practical process for identifying a munition. (See Ordnance Recon.)

Arming Delay: A mechanical, pyrotechnic, or electrical component that delays the arming sequence of a fuze.

Arming Device: An internal component designed to align the explosive train electrically or mechanically within a fuze. (See S&A Device.)

Arming Pin, Safety Pin, Clip, Fork or other Positive Safety: A device inserted into or around a component of a fuze to provide a positive block.

Arming Vane: A propeller-like component that rotates when moving through an airstream or water. (See Water Forces and Wind Forces.)

Arming Wire: A wire attached to an aircraft at one end and a fuze at the other. When the munition is ejected from the aircraft, the arming wire is pulled and the arming sequence begins.

Belleville Spring: A thin, round, concave shaped piece of metal or hard plastic with a hole or firing pin in its center. When the required pressure is applied, the spring "snaps" through its concave center, inverting its shape. Commonly used in anti-personnel landmine fuzes.

Body: The area of a projectile between the bourrelet and the rotating band, or, the primary component of an ordnance item. Oftentimes used interchangeably with the term "warhead."

Bomblet: A munition contained within or deployed from a dispenser. Bomblet usually refers to a munition from an aerial dispenser. (See Submunition and Cluster Bomb.)

Boobytrap (B/T): A device designed to be triggered when disturbed by an unsuspecting victim. When applied to landmines, submunitions, and other ordnance, the device is usually made to appear harmless, and the threat is usually an explosion.

Booster: The largest explosive component of a fuze.

Booster Adapter: A component, containing an explosive booster, used to facilitate a connection.

Bore-Riding Pin: A spring-loaded safety pin held in place with a clip until the munition is loaded, after which it is held in place by the "bore" of the weapon system. Upon firing, the munition exits the bore and the pin is ejected from the fuze, allowing the arming sequence to begin.

Capacitor: An electrical component capable of storing electrical energy for one-time use before requiring a recharge.

Cartridge: Refers to fixed and semifixed projectiles.

Cluster Bomb/Munition: A munition contained within or deployed from a dispenser. (See Bomblet and Submunition.)

Conductor: A material that conducts electricity very well.

Contact: Term associated with underwater mine fuzing.

Creep: The slow "creep-forward" motion of internal fuze components caused by a loss of velocity. Creep can be used to arm or function a fuze. (See Deceleration, Retardation, Set-Forward.)

Creep Spring: A spring used to slow "creep" within components of a fuze during flight.

Deceleration: The gradual loss of velocity after a munition reaches apex of its trajectory. Deceleration can be used to arm or function a fuze. (See Creep, Retardation, Set-Forward.)

Deflagration: A rapid burn that produces intense heat in the form of a fireball; a fuel rich explosion. Propellant and damaged ordnance are capable of deflagrating.

Demilitarization: A munition that has had all hazardous components removed, in which case it should be marked by proper authorities, or a munition subjected to a destructive process that leaves no doubt that the munition no longer poses a threat.

Demining: The process of clearing minefields or areas known to contain landmines.

Density: The mass of a material in a specified volume.

Detent: A mechanical catch that impedes or stops the movement of fuzing components. Detents are used in fuzes to stop or unlock components at specific times during arming sequences.

Detonation: A reaction resulting in a self-sustaining shockwave moving through an explosive material faster than the speed of sound in the reacting material. (See VD and SVD.)

Detonator: Used to initiate a fuze and classified by method of initiation (i.e., percussion, stab, electric, flash. (See Spitback.)

Detonator Safe: A safety feature describing a position the detonator can be in and **not** initiate the munition if it functions.

Direct or Manual Arming: The removal of a safety device, such as a safety pin or clip that partially or completely arms a munition; common on landmines and grenades.

Dispenser: A munition designed to dispense submunitions, bomblets, and cluster bombs/munitions.

Drill Munition: A manufactured inert munition used for training or display. (See Inert and Dummy.)

Dud Fired: A deployed munition that failed to function as designed. (See UXO.)

Dummy: A manufactured inert munition used for training or display. (See Inert and Drill Munition.)

Electrical Impulse: When applied to ordnance—a surge of electrical power provided by a launch platform, an internal generator, or a piezoelectric crystal.

Energy: The ability to do work.

Escapement Device: A mechanical device that regulates the rate that fuzing components are unlocked or moved.

Exploder: A complex fuze used on some underwater ordnance.

Explosive Bellows: Extend when expanding gases from a small explosive charge fill the bellow. These are used to complete electric circuits within some fuzes.

Explosive Remnants of War (ERW): A term used to describe an explosive-filled remnant of war. (See Remnant of War.)

Explosive Train: Alignment of the explosive components in a munition.

Firing Mechanism: A decision-making component present in some fuzes. Usually associated with influence fuzing and other underwater fuzing systems.

Firing Pin: A pin used to strike a primer or detonator in or on a munition.

Flat-Surface Concept: Explosive energy is consistently focused when applied against flat surfaces. (See Munroe Effect in Chapter 1.)

Force: An effect capable of changing the speed or direction of a body in motion.

Fumer: The introduction of gas at a specific rate that fills the partial void created behind a projectile in flight. Disrupting the vacuum and reducing drag greatly increase range.

Function as Designed (FAD): The term used to define when or how a munition functions correctly producing the designed effect.

Gravity: The force ensuring all ordnance will eventually end up on the ground or in the water.

Graze Sensitive: A design feature that affords a fuze the ability to function as designed when a grazing impact is made with a target.

Guide Pin: A fixed pin used to guide a movable component within a fuze.

High-Explosive Dual Purpose (HEDP): An HE munition with a shaped charge.

High-Order Detonation: When a high-explosive (HE) main charge is correctly initiated, reaches its SVD and functions as designed. (See Low Order, SVD, and VD.)

Hydrostatics: Fluid pressure.

Inert: A munition that contains no hazardous components.

Inertia: The forward motion of fuze components caused by the violent deceleration of retardation, or impact with a target. Inertia and creep differ only by the speed at which the ordnance slows.

Inertia Switch: A switch within a fuze that moves upon the violent deceleration of retardation to complete a step in the arming sequence. (See Retardation.)

Influence: A term used to describe magnetic, acoustic, and seismic fuzing.

Insulator: A material with extremely high resistance to electrical conductivity.

Lead: Used to transfer explosive energy from one component in a fuze to the next. (See Relay.)

Low-Order Detonation: Results when an initiated explosive does not reach its potential energy output. Something went wrong, resulting in the SVD not being reached. The explosive may partially detonate, burn, or break up into pieces. Common causes include but are not limited to an understrength or defective initiator, deteriorated or damaged explosives, and improper explosive geometry or chemistry. (See High Order, SVD and VD.)

Luminous Intensity: The amount of light produced by a pyrotechnic composition (usually expressed in "candlepower").

Magnetism: Force exerted by a magnetic field.

Mass: The amount of matter in an object.

Matter: Anything that has mass and occupies space. The three states of matter are solid, liquid, and gas.

Minefield: An area in which landmines or waterborne mines were purposefully deployed.

Motion: The movement of a body. Motion implies that something has moved or changed position. Types of motion include: (1) acceleration and deceleration: change in an object's speed; (2) centrifugal force: forces a body to move away from the center of a rotating object; (3) velocity: distance covered in a given time.

Motion: Newton's First Law of Motion: Every object in a state of uniform motion tends to remain in that state of motion unless an external force is applied to it.

Motion: Newton's Third Law of Motion: For every action, there is an equal and opposite reaction.

Munition: A military ordnance item or single piece of ammunition; used interchangeably with "ordnance." (See Ordnance.)

Ordnance: A military munition or multiple munitions; used interchangeably with "munition." (See Munition.)

Ordnance Recon: The initial approach, observation, examination, and documentation of a recovered munition or munitions. (See Approach and Initial Interrogation.)

Piezoelectric (PE) Crystals: A quartz crystal that produces an electrical impulse when stressed.

Power Cell (Primary): Nonrechargeable power sources, such as galvanic, voltaic, and dry cells.

Power Cell (Secondary): Rechargeable power sources, such as nickel-cadmium and lead acid cells.

Primer: Used to initiate a pyropellent; may also be used to initiate a detonator.

Pyrotechnic Delay: A pyrotechnic train, which after a predetermined time, will initiate components in a fuze; can be used as the primary means of arming or functioning a fuze, and as a secondary means of fuze functioning.

Radome: The outer shell of VT fuzing that encloses an antenna or electronics package. Radomes are usually made of plastic or other materials that do not block or hinder the passage of energy.

Ram Air: An opening that allows high-speed air to enter a component of a munition.

Relay: Used to delay or amplify the transfer of explosive energy from one component in a fuze to the next. (See Lead.)

Remnants of War (RW): A munition or recognizable component of a munition that has been located in or removed from an area of conflict. Also referred to as "war trophies," RW are often UXO or contain other dangerous materials.

Render Safe Procedure (RSP): An EOD procedure used to interrupt the function of a fuze or separate essential components of a munition to prevent it from functioning as designed. Access to US EOD 60-series publications is required to research and determine the appropriate RSP for a munition.

Research: Performed to address Step 7 of the Practical Process for Identifying a Munition.

Retardation: The sudden and violent deceleration from the opening shock of parachute or snakeye fin assemblies, also called retardation devices and high-drag producing fins. Retardation is used to arm some fuzes. (See Deceleration, Set-Forward, Creep.)

Rotor: A disk or ball-shaped component of a fuze that turns or rotates during the arming sequence. Rotors usually contain a detonator.

Safe and Arming (S&A) Device: An arming device designed for more complex fuzing systems, it electrically or mechanically aligns components for proper initiation while safeguarding against unintentional functioning of the munition.

Self-Destruct Feature: A pyrotechnic, electrical, or mechanical feature designed to function a fuze after a predetermined amount of time.

Self-Neutralization Feature: A device or element within a munition that renders it inoperative. It does not make the munition safe to handle.

Semiconductor: Material between the extremes of a conductor and an insulator. Examples include silicon, germanium, and carbon.

Setback: During firing or launch, acceleration causes movable components in a fuze to move backward.

Setback Pin and Spring: A pin held in place by a spring. Upon firing, setback moves the pin rearward, compressing the spring. The fuze arming sequence may use this action to allow other components to move.

Set-Forward: The sudden deceleration from the opening shock of parachute or snakeye fin assemblies. Set-forward can be used to arm or function a fuze. (See Deceleration, Retardation, Creep.)

Slider: A component that "slides" into position during the arming sequence. Sliders usually contain a detonator.

Spin–Centrifugal Force: Fuzing used with spin-stabilized munitions may use the centrifugal force generated to move fuze components outward.

Spitback: A term used to describe a two-section fuzing configuration in which one section initiates the second by "spitting back" an explosive impulse through a tube or void within the munition; commonly used with HEAT and shrapnel munitions.

Stable Velocity of Detonation (SVD): The SVD of an explosive is the speed at which the reaction zone progresses through the explosive without diminishing in strength. Each explosive material has an SVD range based on factors such as its density, temperature, geometry, and method of initiation.

Submunition: A munition contained within or deployed from a dispenser. (See Bomblet and Cluster Bomb.)

Superquick and Delay Selectors: An external component resembling a standard screw that allows a fuze to function in different modes (not present on all fuzes).

Sympathetic Detonation: A detonation caused by the transmission of a shock wave from one explosive, through a medium such as the body of an ordnance item, into a second explosive, resulting in a high-order detonation.

Unexploded Ordnance (UXO): A deployed munition that failed to function as designed. (See Dud Fired.)

Velocity of Detonation (VD): The initial speed at which the reaction zone travels.

Warhead: The section of a munition that contains chemical agents, high explosives, or other energetic materials. This is oftentimes used interchangeably with the term "body," but warhead is more commonly associated with missiles and rockets.

Water Forces: The force exerted by water on a munition or on components as a munition travels during deployment. (See Arming Vane.)

Wind Forces: The force exerted by airflow as a munition travels during deployment. (See Arming Vane.)

Appendix D: Explosives Commonly Used with Ordnance

Each country uses a variety of explosives in the ordnance it manufactures. General information on a few primary, secondary, and main charge explosives often found in ordnance is provided. Information on other explosives is available in many of the references.

Primary Explosives

Primary explosives are the least powerful but most sensitive explosives used in military ordnance. Characteristics include:

a. When exposed to flame, whether confined or unconfined, they will detonate.
b. They are susceptible to initiation from static discharge.
c. They are sensitive to the shock produced from impact.

Examples of primary explosives include:

Lead azide is one of the most common primary explosives in military fuzing and the first-fire mixture in many blasting caps. This white-to buff-colored solid has a Velocity of Detonation (VD) of 17,500 fps. Moderately hygroscopic, it is stable in storage. However, it reacts with copper to form extremely sensitive cupric azide. For this reason, it is usually loaded into aluminum housings.

Lead styphnate is a very common primary explosive in use since WWI. It is used extensively in blasting caps and is easier to initiate than lead azide. It is yellow in color with a VD of 16,900 fps. Slightly hygroscopic, it is stable in storage.

Mercury fulminate was widely used in military ordnance until being replaced by lead azide and lead styphnate. It is yellowish to gray in color with a VD of 17,500 fps. Stability in storage is poor, with a shelf life of less than 10 years. Additionally, it reacts with aluminum, magnesium, copper, zinc, and brass.

Secondary Explosives

These are the most powerful and brisant of the high-explosive groups. They differ from primary explosives in three ways:

a. When exposed to flame, unconfined secondary explosives will usually burn.
b. In bulk, they are not susceptible to initiation from static, but as an explosive dust they are.
c. The shock required for initiation is much greater than primary explosives.

Examples of secondary explosives include:

Trinitrotoluene (TNT) was discovered in Germany in 1863, but not used as a military explosive until the early 1900s. The energetic properties of TNT are the standard by which all other explosives are measured. When characterizing an explosive material, TNT is used as a baseline = 1. It is light brown, tan to slightly yellowish in color with a VD of approximately 22,100 fps. With a melting temperature of 171°F, TNT is easy to use for ordnance applications and a component of many main charges. It is slightly hygroscopic with excellent stability in storage and a shelf life exceeding 50 years.

Trinitrophenylmethylnitramine (Tetryl) is yellow in color with greater brisant and power potential than TNT. Slightly hygroscopic, its stability in storage is dependent on environmental factors, which led the United States to discontinue the use of tetryl in 1979. However, older munitions are likely to contain tetryl boosters in their fuzing system.

Cyclotrimethylenetrinatramine (RDX) was discovered in Germany in 1899, but not used extensively by the military until WWII. White in color with a VD of 28,000 fps, RDX is one of the fastest conventional explosives. RDX is almost 50% more brisant and 75% more powerful than TNT. With a high melting temperature and somewhat sensitive to impact in pure form, RDX is mixed with other explosives or desensitizers for military use. Explosives containing

30% to 98% RDX include Compositions A, B, C, and PE-4. RDX is non-hygroscopic and extremely stable in storage.

Pentaerythritol Tetranitrate (PETN) was developed by the United States and England during WWII. PETN is the most sensitive secondary explosive used in military ordnance. In pure form, PETN can be used as a primary explosive. White in color with a VD of approximately 27,000 fps, PETN is almost 50% more brisant and 70% more powerful than TNT. PETN is nonhygroscopic and in pure form is stored submerged in water due to an inhalation hazard. However, PETN deteriorates when stored at high temperatures.

Main Charge Explosives

Main charge explosives are blended mixtures of secondary explosives and other materials designed to produce specific explosives characteristics. Examples of main change explosives include:

Amatol was developed by the British during WWI to extend TNT supplies. Amatol is a mixture of Ammonium Nitrate (AN) and TNT that can range from an 80/20 to a 40/60 mix. The mixture is reflected in the markings; for example, 80/20 Amatol consists of 80% AN and 20% TNT. Tan to yellowish-brown in color, the VD will vary depending on the mix, but usually ranges between 17,000 and 21,000 fps. Amatol is a very insensitive, but highly hygroscopic. When exposed to water, amatol reacts with copper, brass, and bronze to form sensitive explosive crystals.

Ammonium Picrate (Explosive D) was discovered in 1841, and has been used as a propellant and an explosive. Orange to reddish brown in color with a VD of 23,000 fps, Explosive D is an extremely insensitive explosive. The lack of sensitivity allows it to withstand severe impact, making it ideal for armor-piercing projectiles. Explosive D is moderately hygroscopic with very good stability in storage. However, it will react with lead, iron, and copper to form sensitive explosive crystals.

Composition A (A1, A2, A3, A4, A5, and A6) consists of different formulations of RDX, wax, or other materials containing 86% to 98% RDX. White in color with a VD of approximately 26,000 fps, all of the Composition A explosives are more brisant and powerful than TNT. Composition A is nonhygroscopic and extremely stable in storage.

Composition B: The US definition of Composition B consists of an RDX and TNT mixture ranging from 70/30 to 50/50. Many countries use similar ratios, but simply call it "RDX/TNT" or "TNT/RDX" with the most prominent explosive written first. Tan to brown in color with a VD of approximately 25,000 fps, "Comp B" is slightly more brisant and more powerful than TNT. Composition B or similar mixtures are slightly hygroscopic, extremely stable in storage, and an explosive commonly used in ordnance.

High Blast Explosive (HBX) was developed during WWII and consists of an RDX, TNT, and aluminum mixture ranging from 30% to 40% TNT and RDX, with 15% to 35% of aluminum powder and 5% wax to bind and desensitize. An example of an HBX is "High Blast Explosive mixture 6" (H6), which, due to its explosive characteristics, is used in underwater ordnance. With a brisance equivalent to TNT, H6 is more than 90% more powerful than TNT. Gray in color with a VD of approximately 22,500 fps, HBX is non-hygroscopic, extremely stable in storage, and extensively used where high blast effects are required.

Pentolite was developed during WWII and consists of 49% PETN, 49% TNT, and 2% wax to bind and desensitize. Gray in color with a VD of 24,600 fps, pentolite is slightly more brisant and more powerful than TNT. Pentolite is non-hygroscopic. PETN deteriorates at high temperatures and the presence of TNT further complicates this.

Trinitrophenol (Picric Acid) was used extensively by Japan, Germany, and a number of other countries until the mid-1940s. Yellow in color with a VD of approximately 23,300 fps, picric acid has brisance and power characteristics comparable to those of TNT. For short-term storage picric acid is stable, but over time it will react with most metals and form extremely sensitive explosive crystals.

Tetrytol is a 75/25 to 65/35 mixture of Tetrytol and TNT. A yellowish brown color with a VD of approximately 24,000 fps, tetryol is slightly more brisant and powerful than TNT. Tetrytol is slightly hydroscopic. As with tetryl, stability is dependent on environmental factors and it is no longer used by the United States.

Appendix E: Metric and Standard Conversions

Common Metric Prefixes

Prefix name	Factor	Value
Milli (m)	Thousandth	0.001
Centi (c)	Hundredth	0.01
Deci (d)	Tenth	0.1
	One	1
Deka (da)	Ten	10
Hector (h)	Hundred	100
Kilo (k)	Thousand	1000

Standard Conversions

Standard to metric			Metric to standard		
When you know	Multiply by	To find	When you know	Multiply by	To find
Ounces	28.349	Grams	Grams	0.0353	Ounces
Pounds	0.453	Kilograms	Kilograms	2.205	Pounds
Inches	25.4	Millimeters	Millimeters	0.03937	Inches
Inches	2.54	Centimeters	Centimeters	0.393	Inches
Feet	0.304	Meters	Meters	3.28	Feet
Yards	0.914	Meters	Meters	1.094	Yards
Miles	1.609	Kilometers	Kilometers	0.6215	Miles
Ounces	29.573	Milliliters	Milliliters	0.03	Ounces
Quarts	0.946	Liters	Liters	1.057	Quarts
Gallons	3.785	Liters	Liters	0.2642	Gallons

Sources: NIST Special Publication 1038. May 2006.
TM-43-0001-28. April 28, 1994.

Bibliography

1. Fedoroff, Aaronson, Reese, Sheffield, and Clift. 1960. *Encyclopedia of explosives and related items,* vols. 1–10. U.S. Army Research and Development Command, Warheads, Energetics and Combat Support Center, Picatinny Arsenal, Morris Co., NJ.
2. Meyer, Kohler, Homburg. 2007. *Explosives,* 6th ed. Wiley-VCH VerlagGmbh & Co. KGaA.
3. Urbanski. 1964. *Chemistry and technology of explosives,* vols. 1–4, Department of Technology, Politechnika Warszawa.
4. Thurman, J. T. 2006. *Practical bomb scene investigation.* New York: Taylor & Francis Group.
5. TM-9-1300-214 military explosives. September 1984.
6. FM 5-250 explosives and demolitions. June 15, 1992.
7. TM 43-0001-30 rockets, rocket systems, rocket fuzes, rocket motors. December 1981.
8. TM 9-1370-203-20 unit maintenance manual for military pyrotechnics. January 1995.
9. TM 43-0001-28 Army ammunition data sheets for artillery, ammunition, guns, howitzers, mortars, recoilless rifles, grenade launchers and artillery fuzes.
10. Navy electricity and electronics training series, modules 1–24, September 1998.
11. FM 9-16, explosive ordnance reconnaissance.
12. *Navy explosives handbook,* NSWC MP 88-116, Hall and Holden, Research and Technology Department, October 1988.
13. TM 9-1300-203, ammunition for antiaircraft, tank, antitank, and field artillery weapons, August 1960.
14. *Jane's ammunition handbook.* 1996–1997. Jane's Information Group.
15. *Jane's mines and mine systems.* 1996–1997. Jane's Information Group
16. NAVSEA OP5 vol. 1. Ammunition and explosives ashore.
17. OP 1664, WWI ordnance.
18. McGrath. 2000. Cluster Bombs: Military Effectiveness and Impact on Civilians of Cluster Munitions. London: Landmine Action.
19. TM 43-0001-27, small arms ammunition data sheets.
20. TM 43-0001-29, grenade ammunition data sheets.
21. TM 43-0001-36, landmine data sheets.
22. TM 43-001-37 pyrotechnics data sheets.
23. TM 9-1300-200, ammunition, general.
24. TM 9-1325-200, bombs and bomb components.
25. TM 9-1985-2, German explosive ordnance, bombs, fuzes, rockets, land mines, grenades and igniters. 1953.
26. NAVORD OP 1668, Italian and French explosive ordnance. June 14, 1946.

27. TM 9-1985-4, Japanese explosive ordnance, bombs, fuzes, land mines, grenades, firing devices and sabotage devices. 1953.
28. Information prepared by Combined Material Exploitation Center, RVN, in conjunction with 7AF mobile EOD team. Communist Block projected munitions, fuzes in Vietnam. September 23, 1968.
29. FM 3-23.30, grenades and pyrotechnic signals. September 1, 2000.
30. Forge, J. 2004. The morality of weapons research. *Science and Engineering Ethics* 10 (3): 531–542.
31. AR 385-10.
32. Farragut, L. 1879. *The life of David Glasgow Farragut, first admiral of the United States Navy.* New York: D. Appleton & Company, pp. 416–417.
33. Walters, W. P., and J. A. Zukas. 1989. *Fundamentals of shaped charges.* New York: Wiley.
34. OP 998, aircraft pyrotechnics and accessories, 2nd rev. May 29, 1947.
35. TM 9-1325-200, bombs and bomb components. April 1966.
36. M86 pursuit deterrent munition battery preactivation analysis, technical report ARFSD-TR-92007, US Army, Picatinny Arsenal, Morris Co., NJ. May 1992.
37. FM 5-31, Boobytraps, September 14, 1965.
38. ORDATA, version 2.0, guide to UXO identification, recovery and disposal.

Reference Websites

http://cartridgecollectors.org/?page=introduction-to-artillery-shells-and-shell-casings: International Ammunition Association, Inc. website
http://independent.academia.edu/RaeMcGrath: academic research sharing website
http://maic.jmu.edu/ordata/mission.asp: ORDATA, version 2.0, guide to UXO identification, recovery and disposal
http://world.guns.ru/grenade/: information on shoulder-fired munitions
http://www.civilwarartillery.com/: good reference for US Civil War artillery projectiles
http://www.globalsecurity.org/military/systems/munitions/: global security
http://www.hnsa.org: Naval Ordnance Publications (OP), available on the Historic Naval Ships Association website
http://www.inert-ord.net/: "EJ's ordnance show and tell" offers pictures of WWII era munitions from a number of countries
http://www.pica.army.mil: Picatinny Arsenal, the Joint Center of Excellence for Armaments and Munitions
http://www.scribd.com/doc/28231865/SNC-TEC-Ammunition: Access to US ordnance technical manuals
http://www.usace.army.mil/: US Army Corps of Engineers
http://www.uxoinfo.com: the authority on unexploded technology news and information
http://jeremygregg.com/quotes/issues/landmines

Index

305